Science as a Questioning Process

Science as a Questioning Process

Nigel Sanitt

Institute of Physics Publishing
Bristol and Philadelphia

British Library Cataloguing-in-Publication Data

A catalogue record for this book is available from the British Library.

ISBN 0 7503 0369 7

Library of Congress Cataloging-in-Publication Data

Sanitt, Nigel.
Science as a questioning process / Nigel Sanitt.
p. cm.
Includes bibliographical references and index.
ISBN 0-7503-0369-7 (pbk. : alk. paper)
1. Science--Philosophy. 2. Science--Methodology. I. Title.
Q175.S257 1996
501--dc20 96-15832
CIP

Published by Institute of Physics Publishing, wholly owned by The Institute of Physics, London

Institute of Physics Publishing, Techno House, Redcliffe Way, Bristol BS1 6NX, UK

US Editorial Office: Institute of Physics Publishing, The Public Ledger Building, Suite 1035, 150 South Independence Mall West, Philadelphia, PA 19106, USA

Typeset in TEX using the IOP Bookmaker Macros
Printed in the UK by J W Arrowsmith Ltd, Bristol

To Judith

Contents

Preface

My own background is that of theoretical astrophysics and astronomy, and it is my experiences as a professional researcher that have provided the impetus for the present work. However, it is not idle curiosity or relaxed reflection which has been the guiding spirit, but relentless necessity. To ask questions and havc a thirst for knowledge is a natural characteristic which acts as a prime motivation for those who study science.

In common with most scientists in the UK, my formal education in philosophy was, sadly, negligible. The consequences of this lacuna eventually impinged on me to the extent that I realized that science without philosophy was a pale substitute for the study of deep and important problems which, in particular, face astrophysicists. This work stems from my reaction to this process: a scientist's eye view of philosophy and science.

But why should there be a gap between science and philosophy, or more precisely between science and its philosophy? I have come more and more to realize that the background to this question shows a fundamental flaw in the attitude of scientists to philosophy, a flaw of which I have been in the past just as guilty as anyone else.

As an intellectual pursuit the process of science demands that the rational analysis of what is going on is part of the process itself. This, I believe, is just as true of science as it is of, for example, literature, politics or any other pursuit.

This work represents my journey into the realm of scientific enquiry. That journey was greatly helped by Martin Rees, Cyril Hazard and all the members and staff of The Institute of Astronomy, Cambridge. I thank them for stimulating discussions, helpful advice and hospitality. I thank Wolfson College, Cambridge for allowing me the use of College facilities, the staff of Edgware Library for tracking down many an obscure book and the editors of *Physics World* and *Philosophy Now* for permission to reprint material.

This work would never have got off the ground without the steadfast support of my family: in particular, my wife Judith who typed the manuscript and translated some quotes from French into English, my sons, Adam and Gideon, who assisted me in the preparation of the manuscript, and my daughter Ruth, all of whom put up with me whilst I was writing this book.

Nigel Sanitt

Chapter 1

Introduction

1.1 Presuppositions in science

We all have prejudices and preconceived ideas that permeate our thoughts and actions—no less in science than in other areas of discourse. However, in science, the notion of an objective truth, an ineffable reality has always been a chimeric goal. We can try to avoid this hidden agenda by careful and scrupulous analysis of our theories, thereby expunging these thoughts by naming them. But, for every thought expunged a number take their place—any finite ground gained seems to be on an infinite plain. The problem is that it is impossible to eliminate presuppositions because in order to do so we would have to eliminate ourselves. Even before reading a text, the mere act of picking up a book or journal and looking at the title involves preconceived ideas which may affect the acceptability of the work by the reader just as much as adding meaning to the work.

Further hidden bias may occur, for example, in university course structures, reading lists and even in examination papers. These reflect hidden agenda and collective approaches to a subject, even though they may be out of date; the original template being laid down by a previous generation [1]. This includes the layout and arrangement of material in libraries where, for example, books in well-defined subject areas are more easily locatable than multi- or cross-disciplinary volumes. One particular advantage of the use of computer catalogues in libraries is the ease of finding such material by the use of word-search trawling methods that can locate writings by the presence of a given word in the title or by matching keywords. From the point of view of presuppositions, one spin-off from this is that works are not constrained so much by their subject classification, context with other works or even their shelf position. However, on the other hand no work can be seen completely in isolation. It would be tempting to try to start everything from zero, as Gabriel Marcel advises as a 'harsh necessity', but this just cannot be done [2].

An alternative strategy is to admit the problem, which is of our own making, and try to view science in a way which includes ourselves along with our mental baggage as part of the whole process. Relieved of the need for certainty and accepting science as a process, rather than as an object of study, we can adopt a more pragmatic approach without the need for deductive rigour. As William James puts it [3]:

> *... theories are a man-made language, a conceptual shorthand ... in which we write our reports of nature.*

The problem is of course to avoid solipsism or extreme subjectivism. However, the world (which includes ourselves) is a strong enough legislator to ensure that science is not just a question of make-believe. Both of these approaches have their place and by describing them we underline the choices available to us—although at the expense of a secure understanding. According to Moles [4]:

> *We do not construct what pleases us, but we choose what it pleases us to construct.*

Prejudices occur on different levels. Apart from personal idiosyncrasies we absorb what Feyerabend [5] describes as an 'oral culture' which acts as a 'subliminal commentary on published works', by conditioning the participants as well as providing meaning to material available in print. According to Feyerabend, scientific writing becomes a kind of shorthand. Denuded of meaning in order to give the pretence of objectiveness, it develops into a 'minimal text' which requires an explanatory commentary. Science thus produces texts which like literary texts can be criticized and deconstructed. What is outside the text is important.

Rather than a problem, I see this as an important part of science as an intellectual pursuit and something which it has in common with the humanities. That observations and experiments are theory-laden is inescapable. Furthermore, when theories become embedded in the language of science so that phenomena are 'dressed' by interpretations, they become 'praxis-laden' [6].

Either way, observations, experiments and theories are always inextricably combined with each other in a web of relations.

Objectivity in science becomes like the house in *Alice Through the*

Looking Glass—the more one tries to approach it the further it retreats [7]. Objectivity, at best, is an unattainable and irrelevant fiction and at worst, plainly false if it describes the lack of bias and detachment of a group of seriously committed scientists. Such commitment and single-mindedness provide the backbone to science.

We have not solved the 'problem of induction' and our judgement is hypnotized by the success in mathematics of deductive logic. We look to mathematics, the queen of sciences, for inspiration but when we try to formulate scientific propositions we end up on the shifting sands of nature [8].

The problem is that in mathematics a proposition is capable of being exclusively and immutably true or false. In science any proposition has to be interpreted in the light of a theory, and as such its truth value cannot only change but may not even be strictly definable. Thus, there is a dichotomy between inductive and deductive logic. As noted by Buffon [9]:

> *Truth is different in mathematics from what it is in the natural sciences. In the latter . . . we cannot employ the deductive method of mathematics.*

The 'problem' is that we want to know things in science just as we know that, for example, two plus two make four in mathematics. An easy solution to this problem is to give up and accept that in science truth is something which is not attainable. However, we have this tremendously strong feeling that we do know true things in science and, furthermore, that we know more than we used to know in the past—in other words science progresses. This leads back to the problem of the basis of knowledge in science; if it is not based on truth then on what is it based?

Newton's theory is the supreme example of the changing truth value of a theory following its overthrow by Einstein's theory of relativity. But even within established theories, statements concerning theoretical entities may strictly speaking be neither true nor false and as theories can always be modified to take into account new facts, the truth or falsity of the original theory may thus not be definable.

Because of the enormous success of deductive logic in providing a foundation to mathematics, there has been a feeling that a similar kind of logic can be applied to science. This scenario has failed and, what is more, must fail if it is built on the shaky foundation of the idea of a scientific proposition. On the other hand, the extreme alternative hypothesis that there can never be any kind of logic to science, or that science is somehow haphazard in its operation, does face the evident success of science or at least technological science—we can at least conclude with Shakespeare [10]:

> ... *this madness has method in't.*

The idea of scientific progress is itself problematic and whereas there is a general feeling of the success of science through technological progress, the concept is difficult to quantify with respect to theoretical science. I deal with this point in more detail in Chapter 4.

1.2 Invariance

Central to this problem of scientific propositions as opposed to mathematical propositions is the notion of *invariance*. This term in physics usually has the rather narrow meaning of applying to quantities which are the same before and after an event. I take a slightly broader view and include within the meaning of invariance ideas or concepts which may be abstracted from theories and which display invariant properties. A mathematical proposition is a statement which has a truth value which is invariant in an absolute sense. This is not the case for scientific propositions and it is the inability to pin down the truth value of these which complicates the search for a scientific logic.

The idea of invariance is extremely ancient and fundamental to how we think about the world. The early Ionian philosophers believed that:

> ... *in spite of all the change and transition, there must be something permanent*

although they differed on the nature of this quantity [11].

In his book *An Introduction to Metaphysics* [12], Henri Bergson considers that such a belief in invariable quantities is an ancient error and dates modern science:

> ... *from the day when mobility was set up as an independent reality*

referring to Galileo's experiments with balls rolling down inclined planes.

Bergson was making the point about the importance of the dynamical element of change in physics and the concept of time. Notwithstanding Bergson's remarks, the principle of invariance has been extremely successful over the centuries as a means of obtaining a handle on nature.

Figure 1.1 Henri Bergson. (Reproduced by permission of Mary Evans Picture Library.)

But whether it was water (Thales), air (Anaximenes), fire (Heraclitus), being (Parmenides), the four elements of earth, fire, air and water (Empedocles), nous (Anaxagoras) or atoms (Democritus), the underlying concept of invariance was the same. In later centuries we have energy (Descartes), matter (Bacon), least action (Fermat and Maupertuis) [13] and velocity of light (Einstein) [14]. In Chinese thought in the tenth century there is matter or *chhi* which [15]:

... remains a unity ... and never changes.

Right from the dawn of science and philosophy there was the belief in some deep invariant aspect to an otherwise changing universe. The question posed by Thales is in a sense still being asked and still left unanswered.

It is interesting to note in connection with Einstein's theory of special relativity that he did not originally refer to it by this name but used the term 'invariance theory' [16].

There is much more to the idea of invariance than the modern idea of a conserved quantity [17]. It is more in the nature of an anchor on which to locate and understand the world—for some the invariant is called God. To the more prosaic research scientist, however, it represents a handle on a problem and usually guarantees a mathematically amenable solution; the question is then what are the invariant quantities about which theories can be woven?

Before this question can be addressed we have to ask: by what methods do we choose these invariants and are these methods themselves invariant with respect to the continuous overthrow of scientific doctrines [18]? This potentially infinite hierarchy of theories, each of which guarantees the *truth* of the one lower down in line, underlies the sterility of the concept of inductive truth. In a sense, any logic of science must be reflexive in that it has to allow for self-criticism. It is this criticism which is part of the whole science process.

We cannot escape from the subjective nature of science. Whether you look upon the scientific process as observations of an independent body of knowledge, or as group activity about some objective reality of nature, the subjective element never seems to completely relent. In the interpretation of the facts, there has to be a semantic element—theories have to *explain* and in this way we understand the world. But there is not a straightforward Cartesian duality of the scientist and the world. The compelling seventeenth-century view of science is elegant but sterile, it must be replaced in thought as well as deed. The negative effect of Cartesianism in science is to remove humanity from a world in which twentieth-century science is described in terms of seventeenth-century language. A good example of this is the concept of causality.

1.3 Causality

Inextricably linked with the idea of invariance is that of causality, a concept still heavily tied down by seventeenth-century ideas. (It is surprising how many scientists, at the mention of causality still have this picture of Newtonian billiard balls bouncing off each other.) This rather simplistic view does not take into account the notion of emergent phenomena. What makes water wet or glass transparent? How do you score points in a computer game? These examples are difficult to conceptualize in terms of a limited notion of causality and it is not surprising that in scientific research causality is seldom mentioned. Notable exceptions, however, are in relativity theory in connection with so-called 'non-violation of causality', and in quantum mechanics, so we shall look briefly at these theories.

In a straightforward 'dog bites man' situation you cannot change the past into 'man bites dog'. (*Facta infecta fieri non possunt*: what is done cannot be undone.) Though one must bear in mind the rather pithy aphorism of Nietzsche [19]:

> ... *before the effect one believes in different causes than one does afterward.*

In the language of modern physics closed time-like loops are forbidden. The resulting paradox, referred to as causality violation, as any viewer of *Star Trek* [20] knows, would lead to an illogical world.

We all move forward in time, with the possibility of going back in time and, say, killing your father, before you were conceived, being manifestly impossible, since how could you be here if you were never born? If we move away from the absolute notion of Newtonian time to Einstein's relativity theory, time and space are related by the constancy of the velocity of light in a vacuum. Here, again you cannot travel back into your own past. In relativity a space–time event is an invariant quantity and the space–time manifold is time-oriented so that for each event a distinct choice has been made as to the future and the past, and this choice is continuous from event to event throughout space–time.

If we move to the general theory of relativity, where gravitational effects are introduced and matter and space–time are intimately linked to each other, we might at first think that the model universes generated by the model also legislate against causality violation. Surprisingly, at least to the early workers who studied this problem, this is not the case. This does not, of course, invalidate the theory, but it means that the theory is able to generate models which are mathematically valid but which are disallowed on physical grounds [21, 22].

As in classical physics, invariance also ensures that given the same initial conditions, the results of repeated experiments will be the same and that the absolute position or time is irrelevant [23]. This causal principle that the same cause produces the same effect is entrenched in science as a doctrine and is used in a different sense from the category of causation in general [24]. Echoing Hume, Mario Bunge underlines the fact that 'rigid causal chains exist only in our imagination' [25].

The problem for the causal principle is that even in classical chaotic situations arbitrarily small differences in initial conditions can lead to strikingly different outcomes.

This has given rise to the extremely interesting branch of physics known as chaos theory or chaology. The important aspect of a 'chaotic' situation is that even in strictly Newtonian billiard ball dynamics, outcomes cannot be fully derived from initial conditions [26]. This 'uncertainty' is quite different from the uncertainty of quantum mechanics enshrined in Heisenberg's famous uncertainty principle, which refers to limitations of our simultaneous knowledge of all the parameters of a dynamical system. The amount of novel structure emanating from chaos theory and mathematically tractable by the use of computers, has applications in many other branches of science in addition to physics, as well as giving the playwright Tom Stoppard an interesting theme to one of his plays [27].

It is in quantum mechanics that most of the real problems with causality lie and even though invariance principles abound their aid to scientists may sometimes be unhelpful. For instance there is charge invariance (conservation of charge), conservation of other quantum numbers and PCT (parity, charge, time) invariance [28], but the situation is complicated by the indeterminate nature of reality in a quantum-mechanical system. There are in a sense two realities: firstly, independent reality which is distant and veiled from us and secondly empirical reality which is the totality of phenomena [29]. It is this latter idea of reality with which we are more familiar in our observations of the world.

The quantum state description cannot express information about every observable, so the same reality may be referred to by more than one state description. Even though there may be local consistency, one cannot generally assume that state descriptions are globally consistent pictures of reality [30–32]. Extremely good predictive accuracy for quantum mechanics is of course an antidote to qualms about the validity of the theory itself.

Another aspect of quantum mechanics concerns the indistinguishability of particles. According to Pais [33], the indistinguishability of photons was first suggested in 1911 by Ladislas Natanson who was considering the Planck radiation law. But one must be careful not to confuse indistinguishability with individuality. After all, in classical mechanics we may consider that two objects are indistinguishable from each other but nevertheless are separate individuals, their positions and trajectories being locatable and distinct. From the quantum point of view this is not possible since any labelling of electrons, for example, at one instant would not be carried over to another time. As Bachelard points out, the conceptual problem is that 'spatial localization underlies all language' [34]. Instead of treating the number of particles in a system in the classical sense, we must consider this *number* as a quantum number of the system. In the classical limit this becomes the number of separate particles, but in the quantum limit it does not refer to individual objects but to a property of the system.

The interplay between theory, observation and assumption leads to a hierarchy of divisions—a term used by Van Fraassen [35]. The first two divisions are as follows.

Division 1: elementary particles fall into two classes distinguished by whether the exclusion principle works or not.

Division 2: fermions obey Fermi–Dirac statistics (antisymmetric wavefunctions) and bosons obey Bose–Einstein statistics (symmetric wavefunctions).

These two divisions are neither coincidental nor logically independent and even though they fit very neatly into quantum theory they are to a certain extent independent of it.

The basis of science is therefore flawed, in the sense that the concepts which we have such as reality, causality, objectivity and truth do not give us foundations on which science can be grounded. My aim in this work is to look at science from a point of view that develops a theory or framework within which we can understand the logical rationale underlying the acceptability rather than the truth of theories. The rationale has to be logical and grounded within the world, since we cannot just invent anything we like or adopt any view.

Theories have to be acceptable according to some criteria and in the absence of an objective truth there has to be some utilitarian or practical justification of a theory no matter how abstruse or theoretical the theory may be. To put it another way, a theory reflects in some way a scientist's endeavour to cope with the world, be it the world of a particular set of observations or even of different theoretical ideas which do not fit together. This theme of 'coping' presents science as an integral part of intellectual life rather than as a separate occupation pursued by a small elite, somehow cut off from the rest of humanity. Science is an activity which has developed within mankind just as much as a result of an evolutionary process as other faculties which help us to survive and prosper. I take up this point more fully in Chapters 5 and 6.

My first aim is to look at how theories are constructed and develop the idea in Chapter 2 of the hierarchical nature of theories and their relationship to scientific questions. Out of this it is possible to construct a model of a scientific theory whose underlying form is mathematical, reflecting a logic not of truth values but of combinatorial possibilities. The mathematical details are presented in Chapter 3, and in Chapters 4 and 5 the model is discussed in terms of theory change and the subjective–objective debate.

My second aim is to place theoretical science within the framework of a wider science of humanity and in Chapters 5 and 6 I look at the model from an evolutionary point of view, in particular highlighting an important cross-fertilization with theories of artificial intelligence, statistical analysis and the 'structural' aspects of anthropology and child development psychology.

The next three chapters are devoted to case studies of selected theories. I look at well-established theories in Chapter 7, and in Chapter 8 I have chosen theories which are at the moment of writing fairly speculative. I start with examples from astrophysics, but in Chapter 9, as a contrast, I consider Darwinian theories of evolution. By deliberately choosing theories which are viewed as speculative, even by their adherents, it useful to see how the present model handles theories which may have a short life.

In Chapter 10, I extend the analysis of the model beyond science to literature. At first sight this might seem to be completely irrelevant to the rest of this work, however, my aim is to put science into its proper context and to draw together science and literature as intellectual pursuits commonly rooted in human endeavour. I elaborate this point further in the concluding chapter which draws together the various threads of this work.

Notes and references

[1] Macdonell D 1986 *Theories of Discourse: An Introduction* (Oxford: Blackwell) p 4
[2] Marcel G 1948 *The Philosophy of Existence* (Engl. Transl. M Harari) (London: Harvill) p 93
[3] James W 1949 *Pragmatism* (New York: Longman Green) p 57
[4] Moles A A 1957 *La Création Scientifique* (Genève: Editions René Kister) p 22
[5] Feyerabend P 1987 *Farewell to Reason* (London: Verso)
[6] Heelan P A 1991 Hermeneutical phenomenology and the philosophy of

science *Gadamer and Hermeneutics (Continental Philosophy 4)* ed H J Silverman (London: Routledge) p 226

[7] Carroll L 1871 *Through the Looking Glass and What Alice Found There* (London: Dent) ch 11

[8] Popper K R 1986 *Objective Knowledge* (Oxford: Clarendon) ch 1

[9] Leclerc de Buffon G-L quoted in Copleston F 1985 *A History of Philosophy (Book VI)* (New York: Image, Doubleday) p 52

[10] Shakespeare W *Hamlet* Act II, Scene II

[11] Copleston F 1985 *A History of Philosophy (Book 1)* (New York: Image, Doubleday) p 20

[12] Bergson H 1913 *An Introduction to Metaphysics* (Engl. Transl. T E Hulme) (New York: Macmillan) p 64

[13] Copleston F 1985 *A History of Philosophy (Book 6)* (New York: Image, Doubleday) p 16

[14] Pais A 1982 *Subtle is the Lord* (Oxford: Oxford University Press) p 139

[15] Ronan C A 1978 *The Shorter Science and Civilization in China* vol 1 (Cambridge: Cambridge University Press) p 222

[16] Nozick R 1981 *Philosophical Explanations* (Oxford: Clarendon) p 749 n28

See also Miller A I 1986 *Imagery in Scientific Thought: Creating 20th Century Physics* (Cambridge, MA: MIT Press) p 199

[17] van Fraassen B C 1989 *Laws and Symmetry* (Oxford: Clarendon) p 278

[18] Bromberger S 1969 Science and the forms of ignorance *Observation and Theory in Science* ed E Nagel, S Bromberger and A Grünbaum (Baltimore, MD: Johns Hopkins) p 49

[19] Nietzsche F 1974 *The Gay Science* (New York: Vintage) p 210

[20] *Star Trek* is a science fiction television series.

[21] The Gödel (rotating universe) model allows closed time-like loops. See, for example, Gal-Or B 1981 *Cosmology, Physics and Philosophy* (Berlin: Springer) p 448

Misner C W, Thorne K S and Wheeler J A 1973 *Gravitation* (San Francisco: Freeman) p 922

[22] Pais A 1986 *Inward Bound* (Oxford: Clarendon) p 212. The problem of causality, according to Pais, was raised by Rutherford as early as 1913, although the notion of the lifetime of a particle implies difficulties with classical causality.

[23] Wigner E P 1949 Invariance in physical theory *Proc. Am. Phil. Soc.* **93** 521–6

[24] Bunge M 1979 *Causality and Modern Science* (New York: Dover) p 3

[25] Bunge M 1979 *Causality and Modern Science* (New York: Dover) p 133

[26] Prigogine I and Stengers I 1984 *Order Out of Chaos* (London: Heinemann)

[27] Stoppard T 1993 *Arcadia* (London: Faber and Faber) (First performance at the Royal National Theatre, London, 13 April 1993.)

[28] PCT stands for parity, charge, time. See Davies P C W 1979 *The Forces of Nature* (Cambridge: Cambridge University Press) p 170

[29] D'Espagnat B 1989 *Reality and the Physicists* (Engl. Transl. J C Whitehouse and B D'Espagnat) (Cambridge: Cambridge

University Press) p 7
[30] Redhead M 1987 *Incompleteness, Nonlocality and Realism* (Oxford: Clarendon) p 45
[31] Bohm D 1980 *Wholeness and the Implicate Order* (London: Ark)
[32] Garden R W 1984 *Modern Logic and Quantum Mechanics* (Bristol: Hilger)
[33] Pais A 1986 *Inward Bound* (Oxford: Clarendon) p 283
[34] Bachelard G 1984 *The New Scientific Spirit* (Boston: Beacon) p 126
[35] Van Fraassen B C The problem of indistinguishable particles *Science and Reality* ed J T Cushing, C F Delaney and G M Gutting (Notre Dame, IN: University of Notre Dame Press) pp 153–72

Chapter 2

Theory construction

2.1 Hierarchies and networks

To someone who is learning about science for the first time, its many disparate branches and separate theories must be both a source of comfort and disquiet: comfort, because it enables the student to study different aspects of science separately, but disquiet because there does come eventually a realization that nature does not necessarily pigeonhole its various phenomena in a way that exactly matches different theories and topics of lecture courses. We may start talking about cross-fertilization of ideas and fitting theories together in areas of overlap, but in the end the world is too complicated to approach in such a politic manner [1]. The problem is that in order to make sense out of a seemingly infinite assemblage of data, the natural inclination is to break down the problem into manageable pieces. This harks back to the Aristotelian method of categorization and has been enormously successful through the ages. However, this recipe for action tends to render the unifying aspects of the world more tenuous.

On the other hand, one can go too far in the opposite direction. James [2] warns that:

> *... the unity of things has always been considered more illustrious, as it were, than their variety*

and the quest for unity in everything, just for the sake of it, can be just as illusory.

Regardless of the way in which a theory might be described in a textbook, physical theories are put together in a hierarchical way. Furthermore, a previous understanding of concepts and presuppositions is also involved in the framework of the hierarchy [3]. The problem is

Figure 2.1 William James. (Reproduced by permission of Mary Evans Picture Library.)

that when students learn about theories these are presented in a neat sequential form usually as a sort of quasi-axiomatic set of propositions, each part of which satisfies or solves a carefully selected set of problems.

In contrast to this, at the research level a scientist is usually working on a very small part of a larger problem. How the scientist's work fits into this larger framework can often be fairly invisible to the scientist, whose horizons may be strongly focused within his or her own problem. These two contrasting examples are meant not so much as a criticism of scientists, students or textbook writers, but to show how difficult sometimes it can be to step back and view the 'big picture'.

The historical development of a theory and its current presentation usually bear little resemblance to each other. As a theory passes through its formative period there will be false trails followed and abandoned; some parts will advance further than others only to be later reorganized. Throughout the whole process there is a continual reformulation and reinterpretation of past work. Far from being timeless and objective, the current account of a theory is usually extremely anachronistic and its logic of presentation often owes more to work carried out well after the original work was completed. In this way, the textbook account of a theory represents a kind of quasi-history.

As far back as the end of the seventeenth century Malebranche was espousing what today we would refer to as *connectionism*. Ideas, or traces are imprinted in the brain and their 'mutual linkage is the foundation of all figure of rhetoric' and 'also of an infinity of other things of greater importance ...' [4]. For Malebranche the *cause* of this connection was the 'identity of time when ideas were imprinted in the brain'. The physiological element of mental function is in physical changes in the *fibres* of the brain as a result of the movement of animal spirits caused by excitation of sense organs. Memory, imagination and ideas are produced as a result of mutual excitation of linked traces and the passage of animal spirits through pathways exhibiting less resistance through previous use.

The idea of brain traces is found in other authors: Locke in his 'Essay' [5] distinguishes between the 'waking' and 'sleeping' brain, of which only the former suffers impressions as a result of thought traces. Descartes in his *Treatise on Man* viewed 'idea traces' in more hydraulic terms associating them with the flow of blood [6]. Gassendi talks about 'threads' filled with 'animal spirits' and these same 'threads' appear in a much earlier work published right at the beginning of the seventeenth century [7].

If we translate this into modern language by talking about neurones instead of fibres and electrical activity instead of animal spirits, then Malebranche's account is thoroughly connectionist and is a definite precursor to neural networks [8] (see Chapter 6). Compare this to Minsky's K-lines [9] 'a wirelike structure that attaches itself to whichever mental agents are active when you solve a problem or have a good idea'. For Malebranche the activity of the brain in the thinking process was one of linkages or relationships between different ideas themselves instantiated by neuronal processes.

James, paraphrasing Kant [10], refers to a 'mere motley we have to unify by our wits'. Each concept must then be connected intellectually and *understood* by virtue of its place in the system. The system in question consists of 'parallel manifolds with (each of) their elements standing reciprocally in one-to-one relations'. He discusses the building of a scientific consensus based on three interrelated levels. Disputes that

Figure 2.2 Nicolas de Malebranche. (Reproduced by permission of Mary Evans Picture Library.)

cannot be settled on one level move up to a higher 'scientific court'. The highest level corresponds to the aims of science with scientific methodology one level below and observations and theories inhabiting the lowest level.

For Brown [11] these divisions are not as clear cut, nor are they sufficient in number to reflect the full hierarchy. Watkins [12] employs the step-ladder metaphor with scientific knowledge occupying different levels on a ladder from level 0, corresponding to perceptual reports, to level 4, the abode of universal theories.

But is a hierarchy going too far as a representation of scientific theories, even though hierarchical phenomena in nature are many and varied [13]? Do we necessarily go from one level to the next or are there cross-connections between levels that violate the hierarchical structure, which make us end up going round in circles? The alternative view sees knowledge as a web or a network either with no direction at all [14] or with concepts related by laws [15, 16]. The laws are relations

Figure 2.3 John Locke. (Reproduced by permission of Mary Evans Picture Library.)

of perceived probability or logical coherence, although according to Collins [17] 'they are better described as the networks of social institutions that comprise forms of life'.

According to LeShan and Margenau the use of the term 'hierarchy' can be ambiguous in relation to physical theories [18]. The difficulty is in classifying in hierarchical terms the 'objects' to which a theory refers. LeShan and Margenau use as an example classical physics and quantum mechanics. If size is the determining factor then classical physics can be said to be higher in the hierarchy than quantum mechanics, and if complexity is the important factor then the reverse is the case.

We also have to be careful to recognize that when we infer a hierarchy in a system over time it does not follow that the system is necessarily structured in that way [19]. Such a cognitive hierarchy based on temporal induction 'comes from *us*, and not necessarily from the system that we are describing' [20].

Figure 2.4 René Descartes. (Reproduced by permission of AIP Emilio Segré Visual Archives.)

Searle describes a parallel situation in which he distinguishes between mental states which determine human action and machine information processing which proceeds 'as if' there were mental states [21]. By considering the 'as if' situation purely as a metaphor and not as the way machines really work, he is seeking to weaken the argument that machines can think. 'Human beings' declare Arbib and Hesse [22] 'do not live by rigorous arguments, but rather make decisions that seem plausible, living within a net of elastic entailments and continual interaction with their environment'. Even though the abstract notion of a network or connected structure is plausible as a model for scientific process, one must decide exactly what is being connected.

Laudan [23] sees science as essentially a problem-solving activity [24] and the solved problem as the basic unit of scientific progress. The aim of science is 'to maximize the scope of solved empirical problems while minimizing the scope of anomalous and conceptual problems' [25]. This idea of questions being important and problem-solving being the process of science is useful. However, the 'solved problem'

Figure 2.5 Pierre Gassendi. (Reproduced by permission of Mary Evans Picture Library.)

may imply the 'true proposition' which is of itself problematical. Laudan skilfully avoids this problem but it is still difficult to avoid the conclusion that if a problem is solved then the answer must in some sense be true. Harking back to the example of mathematics, one would like propositions or their scientific equivalent to be the basic building block, but scientific propositions are not invariant. Following the ideas on invariance introduced earlier, I take science not as a problem-*solving* activity but as a problem-*generating* activity.

This, at first sight appears a rather subtle shift of emphasis with minor consequences, but behind it is the concept of scientific questions as the invariant quantities which form the foundations of our understanding of the world. Before we look at the invariant nature of questions we must first consider what is meant by a scientific question.

2.2 Questioning

A fool can raise more questions than seven wise men can answer [26].

Figure 2.6 John Dewey. (Reproduced by permission of Camera Press.)

Asking questions has been part of the human condition since time immemorial. One only has to look at child development and see the incessant curiosity that youngsters display in their quest for knowledge to see that questioning is an integral part of human intellect. As François Jacob puts it in the opening words of the preface to his book *The Logic of Life* [27]:

> *An age or a culture is characterized less by the extent of its knowledge than by the nature of the questions it puts forward.*

Dewey [28] states that 'the most vital and significant factor in supplying the primary material whence suggestion may issue is, without doubt, curiosity'. He takes this as a common theme ranging from a child putting things in its mouth to the intellectual curiosity of science.

For Schopenhauer 'Life presents itself as a problem, a task to be worked out' [29]. Questioning is so much an integral part of our lives that we to a certain extent take it for granted.

Questions have always been important in philosophy [30, 31]. We can look back to Thales at the dawn of philosophy asking about the ultimate nature of the world. As Copleston emphasizes [32] 'the importance of this early thinker lies in the fact that he raised the question ... and not in the answer that he actually gave ...'. A fragment from Heraclitus [33] describes Pythagoras as having 'pursued inquiry further than all other men and, choosing what he liked from these compositions, made a wisdom of his own: much learning, artful knavery'. So the pursuit of knowledge by interrogation was an established principle in ancient times, whether the interrogation was directed at other people or at the world directly.

Aristotle divided questions into four categories: questions of fact, cause, existence and essence. In fact the whole scheme of categories for which Aristotle is best known is based on questions such as: What is it? How big? Of what sort or quality? In relation to what? Where? and When? [34]. In the twelfth century Adam of Balsham [35] stated that 'all discourse starts from question or statement; knowledge of the rules for using these leads to acquisition of ability of discourse in its more complicated forms' and further 'the study of questions must precede the study of statements'. Here we have the recognition that questions are prior to knowledge. Adam of Balsham deliberately used the word 'statements' rather than the word 'propositions' to underline the primal nature of questions.

During ancient and medieval times the normal method of academic and religious discussion was often in the form of *questions* [36] which were used to criticize and originate new ideas [37]. In *On the Eternity of the World* [38] Siger of Brabant states that: 'the first question is whether the human species ... began to exist ... when it had no previous existence whatsoever'.

Compare this to the opening lines of the *Aim and Structure of Physical Theory* by Pierre Duhem [39]:

> *The first question we should face is: What is the aim of a physical theory?*

Both books start with the same three words 'The first question', Duhem thus continues the same 'interrogative' tradition for the pursuit of knowledge.

In the twelfth century, Adelard of Bath wrote a compendium of scientific ideas culled from his travels over the Arabic world [37]. Under the title *Quaestiones Naturales* (natural questions), he detailed his sightings of an earthquake in Syria, a pneumatic experiment in Greece and noted that light travels faster than sound. He did not rely on 'God' as a direct explanation for all natural phenomena, as occurred in later centuries, but only as 'an explanation to be used when all

Figure 2.7 Aristotle. (Reproduced by permission of Mary Evans Picture Library.)

others had been exhausted'. Adelard was also the first English born philosopher, although his family almost certainly originated in France [40, 41].

In the eastern philosophical tradition, the question and answer format was also used as a means of investigation. These dialogues or '*koans*' which originally arose as spontaneous incidents became more formalized during the Sung dynasty (960–1279). They are carefully designed nonsensical riddles which force the student to think at a tangent—or not at all—so that once the solution is found, the *koan* ceases to be paradoxical [42, 43]. Questioning is thus an activity which is so basic to human nature that most of the time it is taken for granted.

The first to propose a logical approach to questions and answers from a relational or combinatorial point of view was Raymond Lull who flourished in the latter part of the thirteenth and early fourteenth centuries. In his *Combinatorial Art*, Lull proposed working

Figure 2.8 Adelard of Bath (Reproduced by permission of The University of Leiden, Scal. 1, f.1^R alleen miniatuur.)

out problems by combining a code of letters of the alphabet in geometrical patterns so as to generate different sets of combinations. The letters represented, in turn, questions, subjects, meanings and relationships. The lists of combinations generated were tabulated by Lull in great detail. However, for Lull his method was much more than just a logical progression but 'a way of finding out and "demonstrating" truth in all departments of knowledge' [44].

Lull was to influence Leibniz who in 1666 published his *Dissertation on the Combinatorial Art*. This was one of Leibniz's earliest works, written at the age of twenty, and was based on the idea of an alphabet of human thought—an idea which first occurred to him whilst he was still at school [45, 46]. Leibniz however, rejected the motivation behind Lull's method which was the conversion of infidels. Leibniz's project was to attain a universal characteristic or symbolic logic which would extract all truths from basic concepts [47]. His view was that all propositions could be reduced to the subject–predicate form and

Figure 2.9 Raymond Lull. (Reproduced by permission of Mary Evans Picture Library.)

that the logic of science was the identification of all true propositions. The method consisted of finding all the possible predicates of a given subject and all the possible subjects of a given predicate. Leibniz's later fundamental works in logic, philosophy and mathematics were very much grounded in these early ideas, which in turn led to modern logic as we know it.

Scientific inquiry is thus grounded at the primitive level by a kind of curiosity [48]. For Heidegger, curiosity is an extension of seeing 'a tendency towards a peculiar way of letting the world be encountered by us in perception' [49]. At the linguistic level it becomes a questioning activity. Curiosity is a much more basic and profound human, as well as animal, drive. This point is explored further in Chapter 6.

It is difficult in our technological age to retrace the conceptual steps that society has taken and start a slightly different path based on more primitive notions. The problem is that attitudes and modes of thought have become entrenched and, in particular, science has become viewed as the study of objective facts rather than as a process in the world. Looking from the point of view of science as a questioning process, there is an important shift in viewpoint. As Randle puts it [50]:

> *The hallmark of the true researcher is an intense curiosity which drives him/her to explore the tangled web of 'why?' and 'what will happen if . . . ' in order to unravel a small portion of it.*

But how are questions related to each other and how do they make a scientific theory? This question will be addressed in the following section.

2.3 Theory as a relationship between questions

Scientific theorizing is seen as a process. The scientist creates his theories both *about* the world and *within* the world. We must avoid the fallacy of imposing a Cartesian subject–object dichotomy on the world. Even a statement as innocuous as 'a theory of science is about the world' might imply by virtue of the preposition 'about' an objective world and a subjective observer. There is, unfortunately, no single preposition that connects 'theories' with 'world' in a way which completely unites and submerges dualist concepts—the French philosopher Renouvier went as far as to say that there is no such thing as science [51].

So how does the *process* of science connect with the problematic or questioning situation and lead to theories of nature? Perhaps an analogy might clarify their connection. Plato used the analogy of a cave to represent the problems faced in the quest for knowledge [52]. My analogy also involves a cave dweller emerging for the first time from his abode, at a primitive stage in the evolution of mankind when language and rational consciousness were being formed. It is raining outside and the cave dweller feels the water spatter onto his face. The physical discomfort invades his conscious awareness and he wonders as to the source of this unfamiliar sensation. At this point, his wonder may not be in any linguistic form—he is like the cartoon character whose speech bubble is filled solely by a single question mark. The situation is represented by the simple animal drive of curiosity. Our caveman attempts to resolve his dilemma—he wipes this strange liquid from his face and examines it. He tastes it, sniffs it, looks at it and in the recesses of his emerging intellect a connection is made. This liquid exactly resembles the water that our troglodyte drank from subterranean

Figure 2.10 Plato. (Reproduced by permission of Mary Evans Picture Library.)

streams. He looks up at the sky as the rain shoots down from the storm clouds, he looks down at the ground as a myriad of rivulets of water infiltrate the sodden ground. A dim awareness sparks in the mind of our cave dweller as to the water cycle of nature. But how does the water fly back into the clouds?

Ignoring the fact that this last question was asked and answered perfectly adequately by Lucretius [53], the point of this story is that even though the caveman cannot answer the last question, the new problematic situation caused by that question does not preclude or negate the progress in understanding what the caveman has already achieved—he has an acceptable theory. All theories have unanswered questions in them. In fact, the whole purpose of a theory is to generate questions. So what is there about this theory of the water cycle that makes it a genuine and acceptable theory?

It is certainly not because we accept the theory as being correct. To say a theory is acceptable because it is correct is to say first that it is true and second we will accept theories which are true. To be acceptable it is a case of conforming to society's norms—or more particularly to those of scientists. But what are the norms and how do they relate to theories? It is because the structure of a theory is valid. The way the caveman's questions and answers are connected enable him to cope with his world.

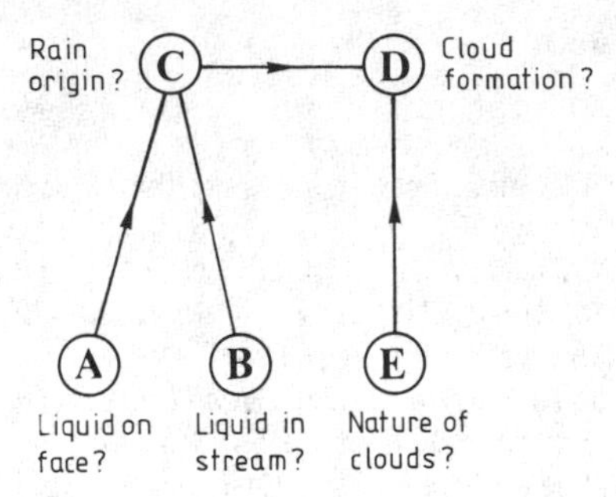

Figure 2.11 Diagram representing the caveman's water cycle questions and answers.

But what is the structure of this connection? The questions 'What is this liquid falling on my face?' (question A), 'What is the liquid in the stream?' (question B), 'Where does the rain come from?' (question C), 'How do clouds form?' (question D) and 'What are clouds made of?' (question E) are all connected with each other as represented graphically in figure 2.11.

The arrows in the diagram represent the answers to each of the questions and indicate the subsequent question raised in the network. The arrows all end on question D—the one unanswered question. The questions A, B and E are what one might refer to as empirical questions derived directly from sense data, and questions C and D are theoretical questions within the scheme; although, today, as opposed to cave-dweller times, these would also be considered empirical questions. The theory of the caveman *answers* three empirical questions by raising two theoretical questions, one of which is unanswered. In this practical sense the theory is acceptable—even if it is later shown to be false.

I have deliberately chosen an example which is at a very basic, almost banal level. It hardly counts as science but because the reasoning involved is so transparent, I am able to rescue its invisibility in order to exhibit the underlying questioning process. It is this very interrogational aspect which I want to emphasize in more complex examples of scientific theories considered in later chapters.

Before we consider what makes theories acceptable and how the connecting structure of questions is created by theorists, we firstly need a formalism to describe in abstract and general terms the key features of the scientific networks we are describing. Such a formalism already exists, in mathematics, and it is to the graph theory that we now turn.

Notes and references

[1] Wigner E P 1949 Invariance in physical theory *Proc. Am. Phil. Soc.* **93** 521

[2] James W 1991 *Pragmatism* (Amherst, NY: Prometheus) p 58

[3] Duhem P 1954 *Aim and Structure of Physical Theory* (Princeton, NJ: Princeton University Press) p 168

[4] Malebranche N 1694 *Search After Truth (Book II)* (Engl. Transl. R Sault) (London: Dunton and Manship) ch 3 p 180
For an up-to-date translation see Olscamp P J and Lennon T M 1980 (Engl. Transl.) (Columbus, OH: Ohio State University Press)

[5] Locke J 1671–1690 *An Essay Concerning Human Understanding (Book II)* ch I, paragraph 15

[6] Descartes R 1992 Treatise on man 177 *The Philosophical Writings of Descartes* vol I (Engl. Transl. J Cottingham, R Stoothoff and D Murdoch) (Cambridge: Cambridge University Press) p 106

[7] Du Laurens A 1603 *Opera Anatomica in Quinque Libros Divisa* (Lyon) (quoted in Brundell B 1987 *Pierre Gassendi: From Aristotelianism to a New Natural Philosophy* (Dordrecht: Reidel) p 94)

[8] Barr A and Feigenbaum E A (ed) 1981 *The Handbook of Artificial Intelligence* vol 1 (Reading, MA: Addison-Wesley)

[9] Minsky M 1987 *The Society of the Mind* (London: Picador) p 82

[10] James W 1991 *Pragmatism* (Amherst, NY: Prometheus) p 76

[11] Brown H I 1988 *Rationality* (London: Routledge) p 234

[12] Watkins J 1984 *Science and Scepticism* (Princeton, NJ: Princeton University Press) p 79

[13] Allen T F H and Starr T B 1982 *Hierarchy* (Chicago, IL: University of Chicago Press)

[14] Sklar L 1985 *Philosophy and Spacetime Physics* (Berkeley, CA: University of California Press) p 169

[15] Hesse M 1974 *The Structure of Scientific Inference* (New York: Macmillan)

[16] Arbib M A and Hesse M B 1974 *The Construction of Reality* (New York: Macmillan) (and 1986 (Cambridge: Cambridge University Press))

[17] Collins H M 1985 *Changing Order* (London: Sage) p 17

[18] LeShan L and Margenau H 1982 *Einstein's Space and Van Gogh's Sky* (New York: Collier Books/Macmillan) p 122

[19] Becker J D 1975 Reflections on the formal description of behaviour *Representation and Understanding* ed D G Bobrow (New York: Academic) p 84

[20] Becker J D 1975 Reflections on the formal description of behaviour *Representation and Understanding* ed D G Bobrow (New York: Academic) p 85

[21] Searle J 1984 Grandmother knew best (Third Reith Lecture first broadcast 21 November 1984) *The Listener* 22 November p 16 and also in *Minds Brains and Science* (London: BBC Publications)

[22] Arbib M A and Hesse M B 1986 *The Construction of Reality* (Cambridge: Cambridge University Press) p 32
[23] Laudan L 1984 *Science and Values* (Berkeley, CA: University of California Press)
[24] Laudan L 1977 *Progress and its Problems* (Berkeley, CA: University of California Press) p 4
[25] Laudan L 1977 *Progress and its Problems* (Berkeley, CA: University of California Press) p 66
[26] Proverb quoted by Telegdi V L 1991 Is quantum mechanics for the birds? *Elementary Particles and the Universe* ed J H Schwarz (Cambridge: Cambridge University Press) p 195
[27] Jacob F 1989 *The Logic of Life* (Baltimore, MD: Penguin)
[28] Dewey J 1910 *How We Think* (London: Heath) p 30
[29] Schopenhauer A 1886 *The World as Will and Idea* vol III (Engl. Transl. R B Haldane and J Kemp) (London: Trübner) p 377
[30] Aristotle 1960 *Posterior Analytics* vol II.I (London: Loeb Classical Library) p 175
[31] Struyker Boudier C E M 1988 *Toward a History of the Question, in Questions and Questioning* ed M Meyer (Berlin: Walter de Gruyter) p 9–35
[32] Copleston F *A History of Philosophy (Book 1)* vol 1 (New York: Doubleday) ch III p 23
[33] Kahn C H 1979 *The Art and Thought of Heraclitus* (Cambridge: Cambridge University Press) p 39
[34] Kahn C H 1978 Questions and categories *Questions* ed H Hiz (Dordrecht: Reidel)
[35] Minio-Paluello L (ed) 1956 *Twelfth Century Logic* vol 1, book 1 (Rome: Storia e Letteratura)
[36] Crombie A C 1961 *Augustine to Galileo* vol II (London: Mercury Books)
[37] Seneca 1972 *Naturales Quaestiones* (London: Loeb Classical Library) a treatise mainly on Earth sciences. Siger of Brabant wrote a number of books on various 'quaestiones', see Copleston F *A History of Philosophy (Book 2)* (New York: Doubleday) p 437 for a list
Adelard of Bath 1920 *Quaestiones Naturales* (Engl. Transl. H Gollanz and H Gollanz) (Oxford: Oxford University Press)
[38] Siger of Brabant 1964 *On the Eternity of the World (De Aeternitate Mundi) (Medieval Philosophical Texts in Translation 16)* (Engl. Transl. and introduction by L H Kendzierski) (Milwaukee, WI: Marquette University Press) p 84
[39] Duhem P 1954 *The Aim and Structure of Physical Theory* (Princeton, NJ: Princeton University Press)
[40] Haskins C H 1927 *The Renaissance of the Twelfth Century* (Cambridge: Cambridge University Press) p 232
[41] Sanitt N 1990 The glorious twelfth *Phys. World* August **3** 64
[42] Kraft K (ed) 1988 *Zen: Tradition and Transition* (London: Rider) p 41
[43] Capra F 1975 *The Tao of Physics* (London: Flamingo) p 56–8
[44] Yates F A 1982 *Lull and Bruno* vol 1 (London: Routledge and Kegan Paul) p 11

See also Johnston M D 1987 *The Spiritual Logic of Ramon Lull* (Oxford: Clarendon)

Gardner M 1958 *Logic Machines and Diagrams* (New York: McGraw-Hill)

[45] Aiton E J 1985 *Leibniz* (Bristol: Hilger)

[46] Parkinson G H R 1966 *Leibniz: Logical Papers* (Oxford: Clarendon) pp xiii, 4

[47] Saw R L 1954 *Leibniz* (Baltimore, MD: Penguin)

See also Parkinson G H R 1966 *Leibniz Logical Papers* (Oxford: Clarendon)

[48] Gale S 1978 A prolegomenon to an interrogative theory of scientific inquiry *Questions* ed H Hiz (Dordrecht: Reidel)

[49] Heidegger M 1988 *Being and Time* (Oxford: Blackwell) p 214

[50] Randle V 1989 Scientists are curious people *Phys. World* **2** 80

[51] Logue W 1993 *Charles Renouvier Philosopher of Liberty* (Louisiana State University Press) p 98

[52] Plato *The Republic VII*

[53] Lucretius 1951 *On the Nature of the Universe* (Engl. Transl. R E Latham) (Baltimore, MD: Penguin) p 235 (Book VI, 609)

Chapter 3

Graph theory

3.1 Introduction

For those not particularly mathematically inclined the first part of this chapter can be skimmed through fairly quickly. The diagram in the previous chapter representing the relations between the questions and answers in the caveman example is referred to as a *graph*, and in this chapter I go through the mathematical theory of such diagrams.

The structural relationships embodied in the diagram and associated theory sets the whole analysis on a mathematical foundation based not on propositional calculus but on a combinatorial algebra.

Graph theory started with a somewhat frivolous problem solved by the Swiss mathematician Leonhard Euler, which was first presented by him as a talk to the St Petersburg Academy in 1735 [1]. Called the Königsberg bridge problem, it involved finding a continuous path over the seven bridges in the city, such that each bridge was crossed just once; Euler proved that it was impossible to find such a route, and in so doing created an important branch of mathematics. He tackled the problem by reducing it to the abstract level of four land areas, represented by points, joined by seven bridges, represented by lines. The *graph* which he produced, consisting only of points and lines, proved to be a mathematical tool of significant importance. Reduced to a more abstract level, the connection with ordinary graphical representation of functions in mathematics can be appreciated. Normally, one would represent y, for example, as a function of x, pictorially as a curve or graph using Cartesian coordinates. In graph theory instead of taking a continuous range of values, x and y form a direct set of integer values. Each value of x may then be represented as a point, and the mapping of x onto y may be represented as a line

Figure 3.1 Leonhard Euler. (Reproduced by permission of Mary Evans Picture Library.)

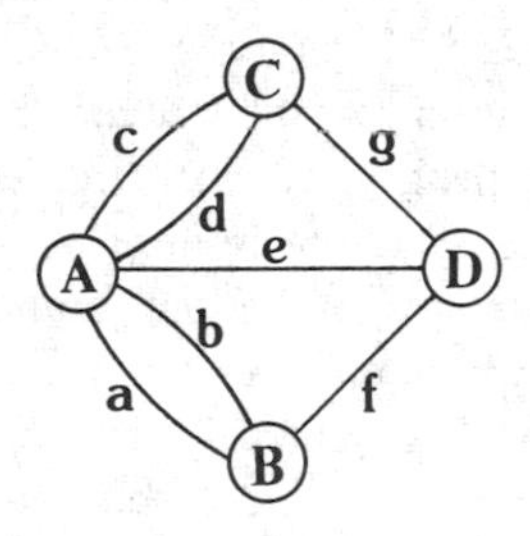

Figure 3.2 Euler graph for the Königsberg bridge problem. The seven bridges are labelled a to g, and the four separate land areas that the bridges span are labelled A to D.

joining each of the two values, x and y. The diagram or graph thus obtained has the same abstract form as Euler's graph which he used to solve the Königsberg bridge problem, see figure 3.2.

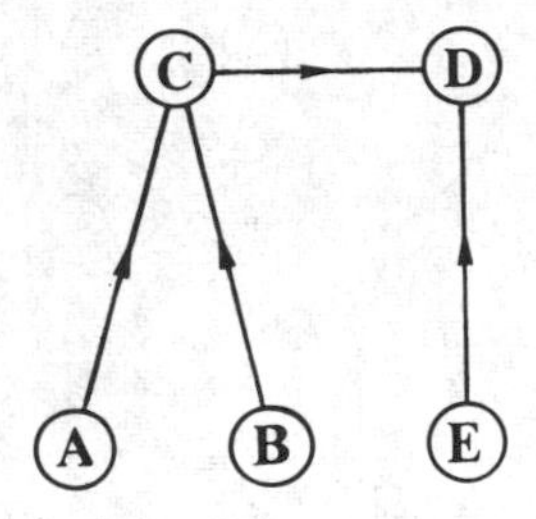

Figure 3.3 Diagram representing the caveman's water cycle questions and answers.

The tremendous power of graph theory lies in its flexibility in that it can be thought of either formally in set-theoretical terms, geometrically as in Euler graphs or even algebraically [2]. For the reader not familiar with this branch of mathematics, one advantage of graph theory is that its objects of study can easily be visualized by the use of graphical examples.

For our purposes, the set theoretical definition of a graph inexorably ties graph theory in as the mathematical formalism to represent scientific theories.

Formally, a *graph* $G = (v, e)$ is a finite non-empty set v of elements called *vertices*, together with a set e of two element subsets of v called edges. Basically, if any sets of objects are connected to each other then they can form a graph.

Recall our example of the caveman creating a theory about the water cycle. His theory consisted of a network of questions as in figure 3.3.

Apart from the arrows on the lines this is, of course, a graphical representation as defined. The arrows mean that the edges between each pair of vertices represent ordered pairs, and such a graph is called a directed graph or digraph. The importance of the arrows in the digraph is that the relations between vertices are ordered, so that for two vertices a and b, a $\rightarrow$ b is not the same as b $\rightarrow$ a.

The edges in a digraph are referred to as *arcs* to denote the fact that they are directed. There are thus two types of graph according to whether edges are directed or not. The word 'digraph' only applies to directed graphs, although the word 'graph' can have both a specific meaning—referring to an undirected graph—or a more general meaning applying to all graphs.

It is not my intention to provide a detailed account of graph theory, since there are numerous textbooks on the subject, many

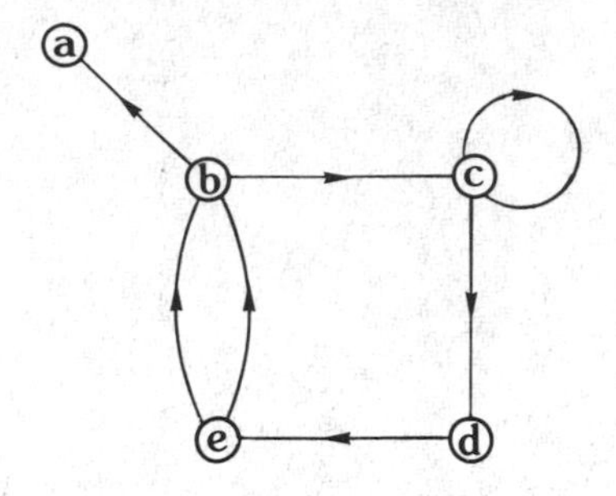

Figure 3.4 Example of a directed graph.

of which concentrate on the varied applications of the theory in such diverse situations as operations research, computer sciences, theoretical chemistry and game theory [3–9]. A brief survey will suffice, concentrating on a few important concepts.

3.2 Terminology

The number of vertices of a graph is called the *order*, and edges joined to a particular vertex are *adjacent* to it. The number of such adjacent edges is called the *degree* of the vertex. In a digraph, an arc may be incident either out of or into a vertex. The number of arcs incident *out of* or *into* a vertex are called respectively the *out-degree* (od) and the *in-degree* (id) of the *initial* or *terminal* vertex. Vertices of degree unity are called *leaf* vertices, other vertices being termed *internal* vertices. The latter term (leaf) has a fairly visual meaning and fits in with the 'tree' analogy which is described later in this chapter.

In the example shown in figure 3.4, there are five vertices and seven arcs. The degree of each vertex are as follows:

$$\begin{array}{ll} \text{id } 1 = 1 & \text{od } 1 = 0 \\ \text{id } 2 = 2 & \text{od } 2 = 2 \\ \text{id } 3 = 2 & \text{od } 3 = 2 \\ \text{id } 4 = 1 & \text{od } 4 = 1 \\ \text{id } 5 = 1 & \text{od } 5 = 2. \end{array}$$

Since every arc is incident to one vertex it follows that the sum of all the in-degrees of a digraph equals the number of arcs. Similarly the sum of all the out-degrees also equals the number of arcs.

A *self-loop* is an arc which starts and finishes on the same vertex. If a digraph has no self-loops then it is termed irreflexive. Two *parallel* arcs have the same initial and terminal vertices. A digraph which has no self-loops or parallel arcs is called a *simple* digraph. The example in figure 3.3 is a simple digraph whereas the example in figure 3.4 shows a self-loop on vertex c and two parallel arcs between vertices e and b.

An *antisymmetric graph* is one in which every arc (t, u) precludes the existence of an arc (u, t), where the initial and final vertices are transposed. The terms antisymmetric and irreflexive are borrowed from the calculus of binary relations which is intimately related to graph theory.

3.2.1 Isomorphism

Two digraphs are isomorphic if there exists a one-to-one correspondence between their vertices which preserves their directed lines. The concept here is one of equivalence or similarity of structure. The concept of isomorphism puts formally in mathematical terms what it means for two graphs to have the same structure. If two digraphs have the same number of vertices and it is possible to order the points so that an arc is in one digraph if and only if the corresponding arc is in the other, then the digraphs are isomorphic. Isomorphism is important when it comes to counting graphs in combinatorial problems. If a given set of vertices can be made to form different graphs given a set of rules, then in order to determine exactly how many different graphs can be so formed all the isomorphically different graphs have to be counted.

3.2.2 Connectivity

In a graph the idea of connectivity is quite straightforward. We can define a *path* as a collection of distinct vertices joined by edges. Thus a graph is *connected* if there exists a path between any two vertices.

In digraphs, there are different kinds of connectivity, because the presence of arrows means that there may be some 'connecting' paths which can only be achieved by going against the direction of an arrow. Thus *strong* connectivity means that for every pair of vertices in a digraph there exists a *directed* path. But supposing we remove the arrows from a digraph (which is not strongly connected) and turn it into a graph. Then, if the underlying graph is connected, the original digraph is said to be *weakly* connected. In figure 3.5 there are two

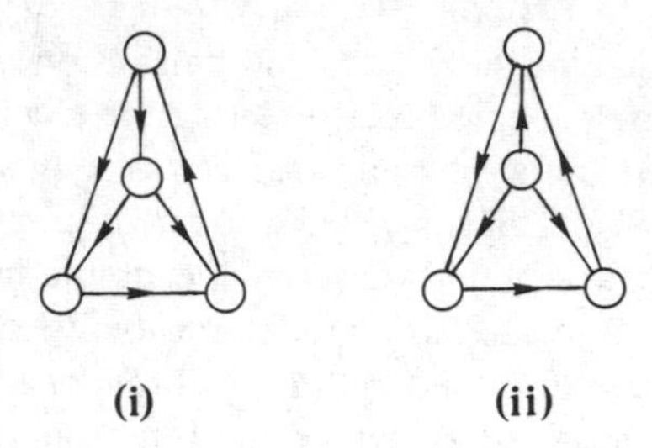

Figure 3.5 Connectedness in digraphs. Both digraphs shown have underlying graphs which are connected, but only (i) is strongly connected.

graphs which both satisfy the criterion for weak connection. However, whereas (i) is strongly connected, this is not the case for (ii), since in the latter case the central vertex is unreachable from any of the outer vertices. (Some writers use the term *walk* to represent the directed path in a digraph.)

3.2.3 Cycles

If an undirected graph contains a closed path, i.e. one in which only the first and last vertices are the same, then such a path is termed a *cycle*. An *acyclic* graph or *tree* is a graph which does not contain any cycles. An acyclic digraph is one which does not contain any cycles, taking into account the direction of the arrows. Figure 3.6 shows an acyclic digraph [9]. Note that in the example shown, the underlying graph is not acyclic. It is this type of digraph which is important for describing the formal properties of relations between questions which exemplify scientific theories.

3.2.4 Condensation

If we remove from a graph a subset of its vertices together with all the arcs incident to and from those vertices we are left with a subgraph generated by the subset of vertices. The condensation, in pictorial terms, is the graph obtained by coalescing vertices and removing loops. It is induced by a partition of the vertex set.

3.2.5 Acyclic digraphs

For acyclic digraphs there are some important properties.

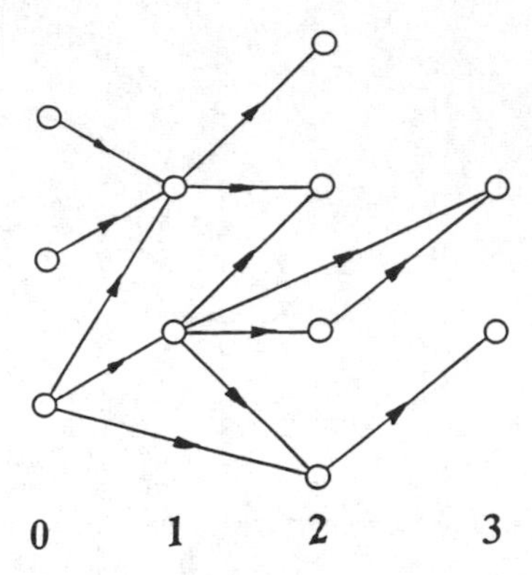

Figure 3.6 Example of an acyclic digraph. The level numbers of the vertices are shown at the base.

Theorem 1. Any acyclic digraph contains at least one vertex which has no successors (sink) and at least one vertex which has no predecessors (source). If we trace a path forward on a digraph, then if every vertex had at least one successor, we would eventually encounter a vertex twice, thereby forming a cycle. Since this is a contradiction, the first part of the theorem is demonstrated. We can prove the second part of the theorem in the same way by tracing a path backwards rather than forwards from a vertex. This theorem leads to the idea of the ordering of all the vertices of such graphs.

3.2.6 Logical numbering and level

A logical numbering on a digraph of p vertices is an assignment i of integers $1, 2, \ldots, p$ to the vertices such that:

(1) each integer is assigned to exactly one vertex and
(2) if (u, t) is an arc then $i(\mathrm{u}) < i(\mathrm{t})$.

Theorem 2. A necessary and sufficient condition for a digraph to be acyclic is that its vertices can be assigned indices in such a way that for all arcs the in-vertex has a higher index than the out-vertex.

A digraph will often have many different logical numberings, but a unique assignment of *levels* can be obtained as follows.

The *levels* function of an acyclic digraph assigns to each vertex the length of the longest path of which it is the last vertex.

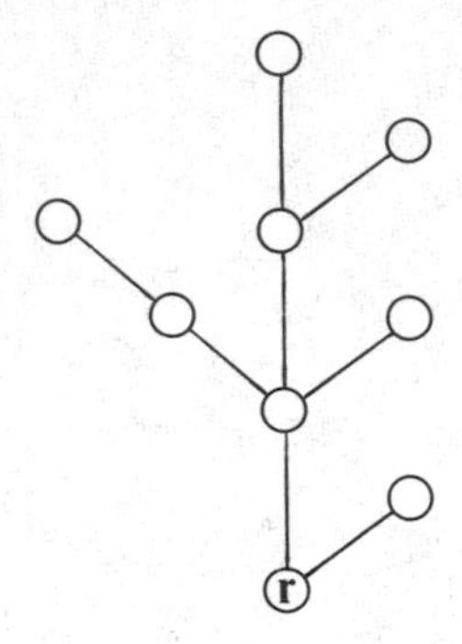

Figure 3.7 Example of a rooted tree. The root vertex is designated r.

The *levels* function has the following properties:

(1) if a vertex u has in-degree zero (source) then level (u) = 0
(2) otherwise, level (u) is one greater than the maximum of the level values of all precursors of u.

The acyclic digraph shown in figure 3.6 displays vertices of the same *level* arranged vertically.

The property of vertex level and the ordering or logical numbering of the vertices in acyclic digraphs formalize the notion of a hierarchy, as discussed in Chapter 2. Thus the common-sense view of relationships is formalized in a very neat way by graph theory.

3.2.7 Trees

The 'tree' image is extremely ancient in all human societies as representational of the universe. Probably the earliest example is the Rosette Tree symbol for Inanna, the fourth millenium BC Sumerian goddess of love (and war).

Trees are the most widely used specialized graph with applications as diverse as genealogy and chemistry [1]. A *tree* is a graph with a designated vertex called a *root* such that there is a unique path from the root to any other vertex in the tree. A tree is connected, since there is a path connecting the root with all other vertices. If a tree is undirected (remove all the arrows), then any vertex can be the root. All the arrows on a rooted tree can be reversed to form a *directional* dual, a rooted tree with all arcs oriented towards instead of away from the root, see figure 3.7.

Theorem 3. For a graph with v vertices and e edges the following propositions are equivalent [2].

(1) The graph is a tree.
(2) The graph is connected and $e = v - 1$.
(3) Every pair of distinct vertices of the graph is joined by a unique path.
(4) The graph is acyclic and $e = v - 1$.

The vertices of degree 1 in a tree are called *leaves*; the remaining vertices are *internal* vertices. The usual terminology is to refer to a directed tree as *rooted*.

In Chapter 2 we considered the example of the caveman's theory on the water cycle (see figure 3.3). This, in graph theory terms, is a tree rooted to the vertex D. We thus have the mathematical model which represents in an abstract way general relationships or connections between entities, but we have now to apply the graph theory model to the networks of questions that we regard as scientific theories.

3.3 Scientific theories

We represent a scientific theory as a network of questions and answers. The questions are the invariant quantities, in the sense that a question, once asked, always retains the same conceptual problematic situation. In the model, the questions are represented by vertices and the connections between questions which are answers, are represented by arcs. I say arcs and not edges, because an answer to a question leads to the next question in a definite and non-reversible sense, so that our graphs must be directed. To say that a scientific theory is a combination of questions and answers, whose ordered linkages may be represented in graph theoretical terms, is to stand on its head the more usual idea of what constitutes a theory.

We usually think in terms of empirical data and basic principles of physics being uniquely blended by intrepid theoreticians into theories. Tentative at first, and then becoming more thoroughly tested by further data aimed specifically at testing the theory's predictive powers to the utmost, a theory finally blossoms to acceptability. Even if it is then modified, abandoned or more usually swallowed by other theories, we tend to think of a theory as describing in a quasi-deductive sense an aspect of nature; as answering questions and providing laws, rather than stimulating questions. In fact the original questions which first gave rise to a theory are often either forgotten or consigned to the backwaters of the history of science. And yet a theory is our conception of how we describe the world and the translation from this to some dehumanized edifice loses so much in the translation that we end up with a false view of science. By concentrating on the questions we can step back behind the stage and see the whole production in a clearer light. By being truer to the spirit of what drives theory creation we can analyse the process and gain a better understanding of science.

3.4 The right questions?

Hintikka [10] sees science as a questioning procedure providing a heuristic pattern to an otherwise unplanned search for truth, the answers generated being synonymous with 'scientific information'. These answers are, in part at least, determined by the questions and the right question will tease out the information required without directly anticipating the answer. But, if we abandon 'truth' in science then there are no 'right' answers and even fewer 'right' questions.

The same argument applies to a 'straight' question which Jardine [11] defines as a question which is:

> ... *both direct and adequate, in the sense that it conveys all the information that is called for.*

Jardine is considering the intelligibility and understanding of questions well posed for a community of inquirers under the term 'locally real'. As Planck [12] pointed out:

> ... *the meaninglessness of a phantom problem is never absolute, but is simply dependent on whether or not a certain theory is accepted as valid.*
>
> This situation can change in time [13]:
>
> ... *the phantom problem of one generation may become the solved problem of the next (and vice versa!).*

There is too much emphasis on answers or possible answers to define the questions or problems posed. Questions are not defined in terms of their *possible* answers but in terms of their *actual* answers to *other* questions which form a network within which the *unanswered* question is embedded.

3.5 Open-endedness of questions

Does an answer to a question always necessarily lead to another question? Why cannot an answer simply neutralize the problematic situation? In a sense this is exactly what happens in a deductive logical system. If I ask, what is one plus one, then the answer—two—does not of itself generate any more questions. If this were the case in science, then the graph theory model would be inappropriate. However, nature is not closed, no answer can ever be the end of the matter because such an answer would be the truth, and no statement in science can be the unquestioned truth. (It does not follow that a repeated train of 'How do you know that?' questions is necessarily meaningful or useful [14].)

The open-endedness of science ensures that theories are always open to be falsified. When this happens the questions still remain, and though some may be dropped and others added to the scheme, a new theory will just be a different structure made of the same building blocks, namely the questions—though the answers may be different. Thus there is a clear difference between mathematical questions whose answers are strictly tautological and scientific questions or problems.

Of course, an answer may include, as part of itself, a deductive element, but this does not impinge on the graph theory model, since all that we see from a semantic point of view is the connection between two questions, any deductive links being invisible on the digraph.

3.6 Acyclic nature of connections

This same open-endedness implies that we cannot have circular arguments in inductive reasoning. A self-loop (in graph terminology) or an answer which implies its own question is a tautology, and therefore applies only to a deductive statement. In the graph theory model the absence of self-loops implies that we have an irreflexive digraph. Furthermore, we can say that scientific theories cannot lead to circular arguments or, in graph theory terms, the digraphs must be acyclic. To see why this must be the case consider a situation where a sequence of questions and answers do form a cycle, shown in figure 3.8 as a subgraph of a digraph.

This would be considered no more than a deductive loop or tautology and the three questions would coalesce to form a condensation within the digraph. The point is that the acyclic character of the graph theory

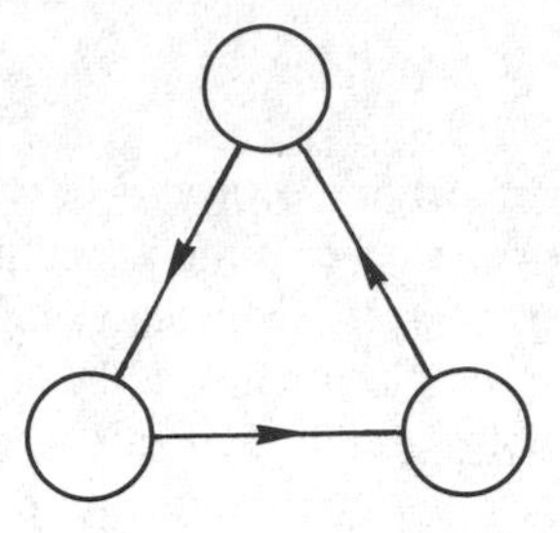

Figure 3.8 Example of a cycle in a digraph.

representation of a scientific theory stems from the inductive logic of the theory; the deductive parts are invisible either within the answers or by the condensation of tautological question and answer sequences. To put it crudely, cycles are not present because, if they were, we would remove them!

To return to the open-endedness of science, Hawking [15] has put forward the idea that fundamental physics may be approaching a conclusion and, in particular, that a 'theory of everything' or TOE is rapidly approaching our grasp.

I expect if one were to ask the troglodyte in the caveman example of Chapter 2 whether his theory about the water cycle was a 'theory of everything' one would receive a positive reply. This is not meant as a tongue-in-cheek criticism; the early Greek thinkers mentioned in Chapter 1 all thought that they had a theory of everything and whether the basis of everything in nature is water (Thales), air (Anaximenes) or fire (Heraclitus) or, in later centuries, God (Descartes, Malebranche), space–time (Einstein) or even TOE (Hawking), I believe the same answer applies to all. Just because one knows that the distance an object falls is proportional to the square of the time taken to fall, it does not follow that the theory which gives rise to that empirical formula is true or cannot be falsified. I choose this example deliberately since the empirical formula was known in the fourteenth century by scholars at Merton College, Oxford [16]. Since that time two major theories of motion have been proposed, namely Newton's and Einstein's, and both of these theories are consistent with the formula (the latter's as an approximation). No doubt, in a few hundred years time objects will still contrive to fall according to the same formula, but will Einstein's theory reign supreme? The answer to this question is irrelevant to this discussion but what is notable is that scientists will quite readily accept that the answer may be in the negative, whereas they would not accept that objects might decide to fall according to some other formula.

Thus, as far as the graph theory model is concerned, scientific theories are represented by acyclic digraphs. As we saw earlier, an important property of acyclic digraphs is that they form a hierarchy, and thus vertices may be assigned in order of level.

3.7 Networks

One major application of digraphs is in the study of network flows. By assigning to each arc a rate of flow of a hypothetical substance, for instance cars or people in a transport problem, then the network obtained can serve as a useful model to solve practical problems. Acyclic digraphs in particular are useful in 'critical path analysis'. This involves the analysis and planning of, for example, construction projects which involve a number of interrelated activities. One can construct an activity graph to represent the project. Each vertex represents the commencement of an activity, the timing constraints being represented by the arcs of the graph. An activity graph is necessarily acyclic, for if a cycle did exist, none of the activities on that cycle could commence. As its name suggests, the principal problem to be solved in a construction project is to identify the critical path, which determines the shortest time in which the entire project can be executed. The question of course in the present applications is why assign a number to an arc, and what would it mean?

An arc represents an answer to a question which links that question to another question. In terms of network flows, what is the hypothetical substance that is flowing? It is the relative importance of the question which is being introduced (by way of an answer) into the structure that determines the number that we may assign to the arc. What are flowing are problem situations, some more important than others, but each linking into the structure and providing its own impetus to inquiry. The network represents a flow of *puzzlement* or to borrow a term from the social study of science we have a network of problematization.

In the context of scientometrics the language content of publications by scientists is analysed in terms of the keyword linkages. The resulting networks are a useful tool to highlight shifting patterns of problems that scientists are studying [17, 18]. This is considered in more detail in Chapter 5.

Borrowing the terminology of network theory, we can identify *sources* and *sinks*. These are vertices which have a net out or inflow of puzzlement.

So far our representation of science is that of an acyclic digraph of questions and answers. This scenario is, however, too general and there are some further restrictions that we must place on the acyclic digraph model.

3.7.1 Vertices with in-degree = 0

These represent questions which have no antecedent answer that connects to them. I use the term 'empirical questions' to characterize these questions. The problem with empirical questions is that the term engenders a split between empirical and theoretical terms. As with all dualisms, the disadvantages of introducing them often outweigh the advantages. From the point of view of the present theory, these empirical questions really represent a cut-off or a black-boxing of theories or presuppositions.

Even in the primitive case of the caveman feeling the water on his face and giving rise to the sequence of questions on the water cycle, the initial question is not really the start of the process. The caveman's brain receives stimulation which is interpreted by him as droplets of water hitting his face. We take this pre-initial sequence for granted, and it is not part of the theory simply because we fully accept this part of the reasoning. We have black-boxed this part of the network and turned it into a deductive sequence, and since deductive entailments are invisible in our theory, the initial question is represented as a solitary question with no antecedents. Instead of empirical and theoretical divisions, all questions should be divided according to whether their vertex had in-degree zero or not. In a sense this represents a convention by which we describe how certain questions are viewed. We perhaps hark back here to the conventionalism of Poincaré [19]. He regarded some aspects of scientific theories as admitting laws by convention; a freedom that a theorist has to adopt certain postulates which experiment has shown to be convenient and which cannot be contradicted by experiment.

Sometimes Poincaré's conventionalism is taken too far. Poincaré himself in his book *The Value of Science* [20] stated that:

> *The scientific fact is only the crude fact translated into a convenient language*

and further [21]

Figure 3.9 Jules-Henri Poincaré. (Reproduced by permission of Mary Evans Picture Library.)

> *The invariant laws are the relations between the crude facts, while the relations between the 'scientific facts' remain always dependent on certain conventions.*

In the present work the 'convention' is to accept certain assumptions or facts as given.

As I discussed in Chapter 1, no theory can ever be entirely free of presuppositions, although there is nothing to stop a black-box being opened up, in which case other presuppositions and black-boxes come into play. The division of questions into empirical and theoretical is a recognition that certain questions have a special status in that we do not consider the prior steps that lead up to them as being part of the theory.

3.8 Theory gain

Just because one can put together a theory, albeit with questions and answers that satisfy all the preceding constraints, it does not follow that what one ends up with is either correct, useful or even acceptable as a theory worth consideration. This is because I have not specified the gain of a theory or what constitutes progress or by what criteria theories may be composed, or even if it is possible to compare theories. There are a number of diverse issues raised here which are discussed in the next chapter, but before we consider these points, I must detail, in terms of the present theory, what constraints on the model incorporate the idea of gain. Looking at the structure of a graph which represents a scientific theory, we have empirical questions or sources which are covered or incorporated into a structure, which also includes other questions which are not themselves empirical. We can thus define the gain as the surplus of empirical questions over other questions which are all included in the digraph or network. The point of a scientific theory is to establish a framework which takes into account as many (empirical) questions as possible. The price which has to be paid is the creation of theoretical structures—in this case questions—which are in a sense one step removed from the world. The balance between these two kinds of questions defines how well a theory has added to our understanding and to what extent there has been a gain which makes the enterprise worthwhile.

In the example used of the caveman and the water cycle in figure 3.3, there are three empirical questions A, B and C, and two other theoretical questions C and D. The gain is therefore one. I can now state the following:

> *Any graph representing a scientific theory must have a positive gain.*

This statement needs a certain amount of elaboration. What I am saying here is that when a scientist argues in favour of a particular theory, either as a lone hypothesis or as an alternative to a rival theory, the underlying basis of a validity argument is the pattern of questions and answers generated by the theory. The more empirical questions encompassed by the theory and the less theoretical structure involved represent an economy of form which renders the theory of superior merit. This is just Ockham's Razor in another guise, but in addition, it is those theories or ideas which just fail to reach this criterion that form the range of sub-theories from which new theories are born. Koestler uses the term *holon* to describe these sub-structures [22]. A theory may be rubbish and it may be wrong but, if nothing else, it must at

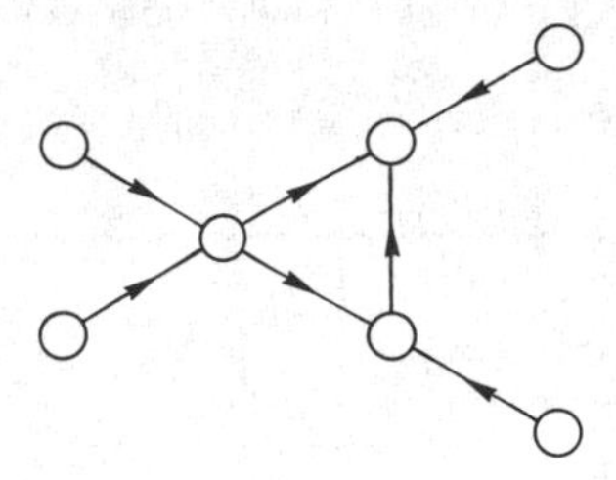

Figure 3.10 Digraph with seven vertices which satisfies the model rules.

least satisfy this criterion at some time for it ever to have been given credence. We can now summarize the rules introduced for the graph theoretic representation of scientific theories as follows.

(1) Theories are represented as acyclic digraphs of questions (vertices) and answers (arcs).

(2) All free-end or leaf vertices have adjacent arcs directed away from the leaf.

(3) The surplus number of vertices having in-degree zero over other vertices must be positive.

We can also introduce a variation on the third rule by assigning numbers to all the empirical vertex arcs and turning the digraph into a network. The numbers represent the relative weighting or importance of the empirical questions. In calculating the surplus instead of the total number of vertices having in-degree zero, we can instead sum the total weights assigned to these vertices and then apply rule 3. The problem with this modification is that the question of what weights to apply is itself problematic.

In a different guise, this sort of weighting goes on already. When theories are proposed or alternative theories discussed, quite often the range of problems and difficulties and their importance are subject to quite a large variation among scientists. Some problems are considered more important than others and not all scientists agree on which have priority.

Applying the above rules 1–3, we can draw an allowed digraph with seven vertices, as shown in figure 3.10.

There are exactly 26 other possible graphs with seven vertices which are isomorphically different from figure 3.10, but which obey the three rules. The number of graphs for different numbers of vertices v and empirical vertices e are shown in table 3.1.

Table 3.1 Table showing the number of isomorphisms for digraphs with v vertices and e empirical vertices, complying with the model.

	v					
e	3	4	5	6	7	Total
2	1					1
3		1	2			3
4			1	3	22	26
5				1	4	5
6					1	1
Total	1	1	3	4	27	

The number of isomorphisms explodes dramatically as v increases. From the above we can say that given, for example, four empirical questions, then there are only 26 different graphs on which the vertices representing these questions can be labelled. The point about the enumeration of possible graphs is that regardless of what answers a theory provides to link a number of empirical questions or problems to each other, in order to be an acceptable theory, there are only a finite number of combinations possible. So that, for example, given three empirical questions—as in the caveman example—there are only three ways in which these questions can be linked together by answers provided by some theory. Furthermore, any alternative theory proposed must have a structure which conforms to one of these three digraphs.

This completes the broad outline of the model which I am proposing: scientific theories are thus seen to conform to this graph theoretic structure or questioning scheme which I describe as a *zetetic* model. The word zetetic means proceeding by inquiry and contrasts with *erotetic* logic which applies to questions in a deductive sense. The word zetetic was first applied to the followers of the Greek sceptic Pyrrho of Elis.

Notes and references

[1] Euler L 1986 *Graph Theory* ed N L Biggs, E K Lloyd and R J Wilson (Oxford: Clarendon) pp 3–8

[2] Gould R 1988 *Graph Theory* (New York: Benjamin/Cummings)

[3] Boffey T B 1982 *Graph Theory in Operations Research* (New York: Macmillan)
[4] Busacker R G and Saaty T L 1965 *Finite Graphs and Networks* (New York: McGraw-Hill)
[5] Harary F and Palmer E M 1973 *Graphical Enumeration* (New York: Academic)
[6] Deo N 1974 *Graph Theory with Applications to Engineering and Computer Science* (Englewood Cliff, NJ: Prentice-Hall)
[7] Harary F, Norman R Z and Cartwright D 1965 *Structural Models* (New York: Wiley)
[8] Robinson D F and Foulds L R 1980 *Digraphs: Theory and Techniques* (London: Gordon and Breach)
[9] Carré B 1979 *Graphs and Networks* (Oxford: Clarendon)
[10] Hintikka J 1981 On the logic of an interrogative model of scientific inquiry *Synthese* **47** 69–83
[11] Jardine N 1991 *The Scenes of Inquiry* (Oxford: Clarendon) p 57
[12] Planck M 1950 *Scientific Autobiography and Other Papers* (London: Williams and Northgate) pp 52–79
[13] Nickles T 1981 What is a problem that we may solve it? *Synthese* **47** 96
[14] Dancy J 1985 *An Introduction to Contemporary Epistemology* (Oxford: Blackwell) p 1
[15] Hawking S W 1980 Is the end in sight for theoretical physics? *Inaugural Lecture to the Lucasian Chair of Mathematics (Cambridge, 1980)* (Reprinted in Boslough J 1989 *Stephen Hawking's Universe* (London: Fontana))
[16] Grant E 1971 *Physical Science in the Middle Ages* (New York: Wiley)
[17] Callon M, Law J and Rip A (ed) 1986 *Mapping the Dynamics of Science and Technology* (New York: Macmillan) ch 7
[18] Latour B 1987 *Science in Action* (Milton Keynes: Open University Press)
[19] Poincaré H 1952 *Science and Hypothesis* (New York: Dover)
[20] Poincaré H 1958 *The Value of Science* (New York: Dover) p 120
[21] Poincaré H 1958 *The Value of Science* (New York: Dover) p 128
[22] Koestler A 1967 *The Ghost in the Machine* (London: Arkana) p 48

Chapter 4

Theory evolution

4.1 Scientific progress

In the previous chapter, I discussed the gain of a theory. Whether in the context of the zetetic model or in general terms, this element can be considered as related to scientific progress. Scientific progress, like the mind, is a concept which is easy to talk about in general terms but is difficult to explain more precisely. In his book on eighteenth century French political philosophy [1], Kingsley Martin anchors the idea of progress firmly in the modern age:

The habit of judging the past and the present by their contribution to the hypothetical future is a child of modern science!

There are two senses of progress. The first is within a theory and refers to an evolutionary process where more and more facts amassed through experiment and observation add to and refine a theory, so that progress represents an encompassing of more and more of the world within a theory or group of theories. The second idea of progress is revolutionary or, in Kuhnian terms, a paradigm shift [2]. Here progress is the complete restructuring of a theory which is overturned by another.

The problem is that, whether viewed as evolution or revolution, change does not equal progress. This is just as true in theoretical physics as it is in life in general. Whereas we may all agree that things are changing, the idea of progress is still problematic.

Many books and periodicals contain detailed writings on scientific topics and an increasing amount of technological gadgetry impinges on our daily lives. We can look at the tremendous increase in computer power or successful rocket launchings throwing ever more satellites into orbit around the planet. Here seems to be concrete evidence of scientific progress, but is it?—and what about theoretical physics? Is progress at the practical level synonymous with progress at the theoretical level?

I believe the answer to this is no. We can see why by looking at an example of enormous practical success coupled with little theoretical progress, namely the use of anaesthetics in medicine. At a practical level the use of anaesthetics is of far-reaching benefit to mankind, yet, at a theoretical level, the actual mechanism which causes loss of consciousness is not well understood. In this case, one may argue that as anesthetics can be used effectively and safely, it does not matter if there is no precise theory as to how they work.

However, in physics, the situation is quite different—empirical laws, no matter how reliable, are no substitute for a theoretical understanding. As Collingwood states uncompromisingly [3]:

> *A scientist who has never philosophized about his science can never be more than a second-hand, imitative, journeyman scientist!*

Of course, experiment and practical experience are important and Collingwood ruefully continues:

> *A man who has never enjoyed a certain type of experience cannot reflect upon it; a philosopher who has never studied and worked at natural science cannot philosophize about it without making a fool of himself!*

However, it is clear that progress at the level of theoretical understanding is the most important and the most difficult to quantify.

4.2 Scientific explanation

Progress is thus seen in the context of increasing explanatory power although there is much more to an explanation of the world than a particular theory. As Nietzsche warned [4]:

> *It is perhaps dawning on five or six minds that physics, too, is only an interpretation and arrangement of the world (according to our own requirements, if I may say so!) and not an explanation of the world.*

Rejecting an 'ultimate' explanation, Popper views the whole process as a hierarchy [5]:

> *Every explanation may be further explained, by a theory or conjecture of a higher degree of universality,*

Figure 4.1 Karl Popper. (Reproduced by permission of Camera Press.)

and further:

> *There can be no explanation which is not in need of a further explanation.*

We have a network of relationships comprising theories with associated mathematics, embedded within further relationships of experiment and observation, other theories, scientific methodology, and social and cultural criteria etc: explanation ends up as pattern generation [6].

Philosophical reflection is not synonymous with theory creation, but steps back and looks over the shoulder of the scientist, examining the framework of the theory that is created. A scientist—Collingwood's journeyman scientist—may be performing an experiment using some complicated looking piece of equipment, which entails adjusting various dials and monitoring a number of meter outputs. When asked about the philosophical relevance of all this, the scientist may not consider that what is being performed has much relevance to philosophy or vice versa. But if the questioning process is pressed further and the scientist is forced to explain what is being done and why, then there very quickly comes a point when discussion is firmly within the province of the philosophy of science. The surprising thing is that in spite of the ease and rapidity by which such lines of thought lead to philosophizing, many scientists fail to follow such paths for most of their working lives. This is probably just as true in our own times as it was in Collingwood's.

In his autobiography, Collingwood emphasizes his ideas on the questioning activity in knowledge [7]. Knowledge for Collingwood is 'both the activity of knowing and what is known' and suggests that 'a logic in which the answers are attended to and the questions neglected is a false logic!' He relates the meaning of a proposition to the question it answers and, further, he sees 'truth' as only something that applies 'to a complex consisting of questions and answers' not to an individual proposition. This idea of looking away from propositions to groups or complexes of questions and answers is the foundation of the present work.

Collingwood's ideas are echoed by Gadamer [8], who discusses hermeneutics as follows [9]:

> *One of the more fertile insights of modern hermeneutics is that every statement has to be seen as a response to a question and that the only way to understand a statement is to get hold of the question to which the statement is an answer. This prior question has its own direction of meaning and is by no means to be gotten hold of through a network of background motivations but rather in reaching out to the broader contexts of meaning encompassed by the question and deposited in the statement.*

For Popper [10] it is problem situations which are at the nub of physics. 'Every solution of a problem raises new unsolved problems' [11]. He visualizes science as progressing from problem to problem, thus 'the most lasting contribution' to the growth of scientific knowledge that a theory can make is the new problems which it raises [12]. Progress is the creation of new problems by constant criticism of theories which leads to error elimination and further problems along a never-ending chain [13]. He represents his scheme pictorially as shown in figure 4.2.

In figure 4.2, P1 and P2 are different problems; TS stands for tentative solutions and EE stands for error elimination. However, in this scheme there is a multiplicity of tentative solutions and the full

$$P_1 \rightarrow TS \rightarrow EE \rightarrow P_2$$

Figure 4.2 Popper's pictorial representation for problems (P), tentative solutions (TS) and error elimination (EE).

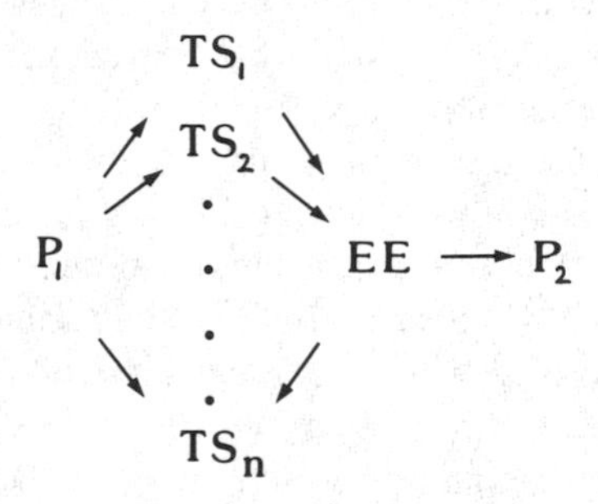

Figure 4.3 Same as figure 4.2, but an iterative procedure generates *n* tentative solutions.

diagrammatic representation is as shown in figure 4.3.

This evolution model of plastic controls applies equally to the evolution of organisms or of theoretical physics. At the scientific level there is 'conscious criticism under the regulative idea of the search for truth' [14].

Einstein also represented his philosophy of science by way of a graphical sketch in a similar vein [15].

Competing theories are evaluated critically and in their destruction new problems are raised. The advance achievement can be assessed by the intellectual gap between the original problem and the new problem, which results from the breakdown of the theory. For Popper 'the learning-process is not a repetitive or a cumulative process but one of error-elimination'. Animals, and even plants, are problem solvers. And they solve their problems by the method of competitive tentative solutions and the elimination of error [16]. Popper takes the biological metaphor for the evolution of theories even further into the realm of epistemology and talks about the growing tree of knowledge [17]. This tree of pure knowledge contrasts with the evolutionary tree of living organisms or applied knowledge. The latter has a main stem which forms into branches representing more specialized or differentiated forms which are so integrated that 'it can solve its particular difficulties, its problems of survival'. The tree of pure knowledge, on the other hand, is represented 'as springing from countless roots which grow up into air' and which 'tend to unite into one common stem'. Popper sees this integrative growth as a result of our ability to criticize through the medium of language. The tree metaphor is part of an ancient tradition as referred to in Chapter 3 and was also the basis of the 'combinatorial art' of Raymond Lull already mentioned in Chapter 2.

The methodology of science as dealing with problem situations is one which tries to converge on or aim at *truth*. Truth is the ultimate yardstick for progress. This truth-oriented or truth-seeking rationale is the point of departure for Laudan [18]. He highlights the problem-solving approach to science, but treats questions of truth as irrelevant. Truth in science does seem to engender its own problems. Kuhn in his book *The Structure of Scientific Revolutions* [19] managed to actually avoid the word until the last few pages, apart from a passing reference to Bacon. He used it just four pages from the end in his final chapter on progress.

According to Laudan [20]:

> ... *the solved problem is the basic unit of scientific progress, the aim of science is to maximize the scope of solved empirical problems while minimizing the scope of anomalous or conceptual problems.*

An empirical problem in his scheme is 'anything about the natural world which strikes us as odd, or otherwise in need of explanation' [21]. 'A conceptual problem is a problem *exhibited by some theory or other*' and is a 'higher order question' [22]. The path of progress is one which '*shows an increasing degree of problem-solving effectiveness*' [23].

The problem with this definition of progress is that to say a particular problem has been solved necessarily encompasses the concept of truth that Laudan declared irrelevant to start with, because as Sarkar points out [24] 'problem-solvers ... can never know anything for certain'. A scientist who 'thought he had solved a problem ... could always be mistaken in principle'. Furthermore, 'there is no formal model in which theories entail problems' and the 'only viable analysis of entailment is in terms of logical consequence, which in turn essentially rests on the notion of truth'. Also, a consequence of Laudan's view is that '*it allows for the possibility of claiming that any false theory, and one known to be false, may be regarded as a problem-solver*' [25].

The problem basically is in the word *solution*. From a conceptual or semantic point of view the solution to a problem represents a closure, an end to the search for a solution. Instead of being *an* answer, a solution represents *the* answer to a problem. As soon as a question or problem is solved the implication is that there is an entailment involved in going from the question to the answer and this in turn implies a truth value to the solution. Problem-solving thus ceases to be open-ended and the inductive argument is closed, implying that the solution is a deductive truth.

In the zetetic model the logic is one of problem-generation with the open-endedness enshrined within the combinatorial rules that govern the linkage of questions. Questions are the only invariant quantities, the answers can be amended, substituted or changed by the scientist in the light of experiments performed or observations made within the whole panoply of science. Provided that the answers make sense in terms of our view of nature, then a valid theoretical model is formed.

Of course, the actual interrogative sentences which we use (and in whatever language) are separate to the questions themselves, or *interrogabilia* as they are termed [26]. Truth does not appear, overtly or covertly, because problems are never solved but are simply linked together by answers which are downgraded to tenuous links which may be easily broken or reformed. The model copes with problems rather than solving them. There is thus no deduction, no truth.

This is not to say that scientists may not believe a theory to be true. It is just that the concept is not one which can apply to a theory. Validity is a question of combinatorial structure permitted by a set of rules. These rules are not themselves true or false, but simply define how we, as humans, view the world as scientists [27].

In a later book [28], Laudan sees progress as relative to a certain goal state [29]:

> *Judgments of progress are ... parasitic on the specification of goals. If scientists move closer to realizing or achieving this state then progress has been made.*

Laudan emphasizes that these goals are not synonymous with the goals of the scientists who originally created their theories. Laudan sees 'no escape from the fact that determinations of progress must be relativized to a certain set of ends, and that there is no uniquely appropriate set of these ends' [30].

In this way, Laudan opens out the model, removing the closure of the word 'solution' and the implication of a truth function.

In the zetetic model, the idea of truth is lost and the formal aspects of the model are in the network of questions which are related to each other. This is fairly close to a coherence theory of truth which according to Ajdukiewicz is 'the agreement of thoughts among themselves' [31] or following Joachim's description [32]:

> *... truth is linked to truth until the arrangement constitutes that network of chains of truths which is the system of ideally complete knowledge.*

However, it is not truths or answers that are being linked together, but questions. This avoids the problem of defining truths in science—which I believe is not possible anyway—and instead relies on questions to form the grounds on which science is based.

Popper [33], considered Newton's and Einstein's theories and suggested a criterion of comparability, or measure of theory content as the ability of one theory to answer questions which the other cannot answer either as precisely or at all.

Watkins [34], refers to this as a *question-answering criterion* and highlights a number of serious difficulties and logical inconsistencies in applying this criterion. The problems, at root, lie in the deductive assertions of the 'answers' and the assumption that theories represent a structure of true propositions. The criterion or what Grünbaum calls 'erotetic comparability' [35] is thus doomed from the start.

The present criterion of 'zetetic comparability' does not involve the comparison of true answers and furthermore is not concerned with the content of a theory so much as its structure.

4.3 Theory change and comparison

In the zetetic model, presented here, theory change is represented by amendments, deletions or additions to a diagraphical construct of questions. The problem of theory change and comparison has been highlighted by Feyerabend, Kuhn, Lakatos and others.

Kuhn's model of paradigm shifts [36] draws particular attention to radical changes amongst scientists' conceptions as a result of a '*Gestalt*' switch to a major alternative theory. The change may be so radical both in terms of conceptions and presuppositions that comparison between the two theories is problematic. Both Kuhn and Feyerabend use the term incommensurability to describe the problem. For Feyerabend [37]:

> *... the replacement of one comprehensive theory by another involves losses as well as gains.*

The problem includes 'meaning invariance' where concepts such as mass undergo a change in meaning in their employment in different theories. Falsification, in the Popperian sense, then becomes a problem since 'an observation statement purportedly corroborating one theory while falsifying another does not actually do so, since, given meaning variance, the terms it employs would not have the same meanings in both theories' [38].

Looking at a theory in terms of theoretical concepts generated leads to problems for the meaning of those concepts when the theory runs into trouble. The problem is that the level of 'concepts' is too high up in the hierarchy of a theory to be salvageable on the theory's demise or major restructuring. By concentrating instead on the questions within a theory, we are at a much more basic level, and the problem of meaning variance is more easily understood.

Incommensurability represents an incompatibility or competition between differing points of view [39]. In the zetetic model, because questions are viewed as the invariant quantities, the problem becomes one of supporting this very contention. As Simon points out [40]:

> *The view that question asking rather than question answering is the critical part of the creative forces would be hard to defend in its extreme form ... in implying a sharp boundary between question asking and question answering, it may be passing the wrong question.*

and further the

> *... reformulation of questions—more generally, modification of representations—is one of the problem-solving processes.*

The point here is that if questions are invariant then how can questions be reformulated—a process which undoubtedly takes place all the time in science? In order to resolve this apparent discord, I now turn to the 'black-box'.

4.4 Question reformulation and 'black-boxing'

That questions are reformulated in the light of experience, experiments or theories is heavily entrenched in our everyday experience. But this reformulation does not alter, in an important sense, questions which would violate the invariance I had postulated. Rather, these questions have simply been 'black-boxed'. I previously described black-boxing as the replacement of one section of the network of questions by a single question. The hidden sequence of original questions is no longer visible because the whole matrix has been black-boxed, or as Baudrillard so elegantly puts it [41]:

> . . . *where all the commandments, all the answers ferment.*

Of course, there is nothing to stop a theorist delving into the black-box but he would have little reason to do so, since the black-box represents an area which is regarded as unproblematic and where there is universal consensus among scientists.

That is not, of course, to say that this does not happen, after all most great advances in science spring from scientists who question foundations in this way. But for the mainstream scientific process black-boxes remain sealed.

There are two points here. First, it is clear that what I refer to as 'black-boxing' of a group of questions does not involve an alteration of a question or disturb its invariant characteristic. A question, once posed, and which thus represents a problematic situation, cannot be unposed. A question can, of course, be dropped—for being no longer relevant—or augmented or black-boxed with other questions. But whatever happens the original question does not lose its invariant quality.

My second point is that whereas most of the time black-boxes are not opened up, there will be occasions where this is not the case. In these cases the original questions which form the black-box become fused in a new network, thus forming new theories and new black-boxes within them.

Given a question and its subsequent reformulation, there has to be an underlying theory or set of presuppositions including auxiliary conditions through which the reformulation is enacted. The reformulation is then the replacement of the original question together with associated questions by a black-box consisting of a single question which now replaces the original question. It is in this sense that the original question has the invariance property. One can always also 'open the black-box and inspect its inner workings' [42], and if necessary create a new structure. This, however, does not alter the invariant nature of the constituent questions.

4.5 Demarcation between questions and answers

In the zetetic model the role of answers has been downgraded to that of providing links between questions which may be broken and reforged in line with the scientist's point of view of the world. But there is an important asymmetry in the model—*all answers lead to questions, but not all questions lead to answers.*

No answer can close a line of thought unless it is a deductive answer, which our model excludes, or more accurately renders invisible. Any answer must give rise at least to the question: why is this particular answer valid? Only a purely deductive answer admits no further questions, since tautologically speaking the only reference is to the original question. Any theory always has unanswered questions within it. It follows then, that any putative answer to a question within the framework of a theory must always lead to another question or questions.

The first part of the asymmetry is represented in the graphical model by the condition that all arcs have vertices at both ends. Thus all answers both spring from and lead to questions. The second part of the asymmetry follows from Theorem 1, Chapter 3, i.e. any acyclic digraph has at least one vertex which has no successors.

The demarcation between questions and answers can be approached considering an answer and its subsequent question, as opposed to a question and its subsequent answer. It is very difficult to move away from everyday language and terminology which is subject to our normal thought processes. If I say 'consider a question and answer sequence' then one thinks of examples of the type 'What day is it today? Today is Sunday' or 'What is the capital of England? It is London'. If on the other hand I say 'consider an answer/question sequence' then my meaning would probably be misinterpreted. In the example of the caveman I could pose the following answers/questions:

(1) Water falls from the sky, where does it come from?
(2) Water is in the stream, where does it come from?
(3) Clouds produce rain, how do they form?
(4) Clouds are in the sky, how do they form?

The essential part of these answer/question pairs is 'the question'—the answer, now in an antecedent position, merely supplying the linkage to the previous answer/question pair. Rather than looking forward from a question to an answer, we go from one question to another using answers as links. Clumsy as this formulation seems from the everyday language point of view, it puts the complex mesh of problems into a form which can be more easily combined into a theoretical structure. Within the graph theory model, we can think of putting together a theory framework by combining together groups of answers/questions.

4.6 Empirical questions

Another demarcation which has proved impossible to establish is the empirical–theoretical division. General criteria may be used to place terms along a spectrum whose extreme ends are observational and theoretical [43], for example reliance on instruments. However, the theory that underlies observations and the empirical content of theoretical quantities deny credibility to this division however compelling it may appear. It is very easy to fall into the trap of thinking that a simple act of observation, such as measuring a length using a ruler or timing an interval using a stop-watch, has no theoretical background. Yet there is in both cases a black-boxing of causal links—a network of questions in the zetetic model—which are so close to being accepted as deductive links, that they are rendered invisible. The empirical distinction then falls back on the observer, who regards an observation as empirical not through any special property of nature with regard to the observations, but because the network of theoretical antecedents is accepted as unimpeachable.

Thus in the zetetic model questions are empirical—represented by vertices with in-degree zero—by virtue of the way that scientists look at nature. The compelling nature of the distinction is therefore a result of our interaction with nature and the way we model the universe. The empirical–theoretical division as presently defined is thus one of a qualified convention. I use the term ‘qualified’ since the choice is not entirely arbitrary, but depends on how a scientist or a group of scientists view the whole theory that is under study. Empirical questions (as defined) are the inputs or genesis of a theory and are empirical simply by virtue of the fact that the questions and answers which lead to them are made invisible by the theory builders. Their history or causal antecedents are detached and absorbed so that they are an interrogation which has become ‘given’, at a very basic level—poetically described by Bachelard as a perceptual image which reverberates [44]. The empirical–theoretical division is thus transformed into one of distinguishing genetic questions.

The invariant nature of scientific questions is thus seen in terms of an openness of questions and the fact that they defy deductive closure by their very nature. In the realm of philosophy itself, such open-endedness is so obvious that it is taken for granted. After more than two millennia most philosophical questions are still unanswered, the word 'still' being superfluous, since no-one expects any definitive answers in philosophy in the millennia to come. And yet somehow in science there is still the belief that around the next corner all will be revealed. It is thus in the inductive, rather than deductive, nature of scientific questions that a failure to definitely answer a question leads to questions having an autonomy which renders them invariant.

Notes and references

[1] Martin K 1962 *French Liberal Thought in the Eighteenth Century* ed J P Mayer (London: Phoenix House) p 279

[2] Kuhn T S 1970 *The Structure of Scientific Revolutions* (Chicago, IL: University of Chicago Press)

[3] Collingwood R G 1945 *The Idea of Nature* (Oxford: Clarendon) (and 1960 (Oxford: Oxford University Press) p 2)

[4] Nietzsche F 1990 *Beyond Good and Evil* (Engl. Transl. R J Hollingdale) (Baltimore, MD: Penguin) p 44

[5] Popper K R 1972 *Objective Knowledge* (Oxford: Clarendon) p 195

[6] Salmon W C 1984 *Scientific Explanation and the Causal Structure of the World* (Princeton, NJ: Princeton University Press) p 121

[7] Collingwood R G 1939 *An Autobiography* (Oxford: Oxford University Press) ch V (paperback 1970)

[8] Gadamer H-G 1981 *Reason in the Age of Science* (Cambridge, MA: MIT Press) p 45

[9] Gadamer H-G 1981 *Reason in the Age of Science* (Cambridge, MA: MIT Press) p 106

[10] Popper K R 1982 *Quantum Theory and the Schism in Physics* (London: Hutchinson) p 160

[11] Popper K R 1983 Knowledge without authority *A Pocket Popper* ed D Miller (London: Fontana) p 55

[12] Popper K R 1983 The growth of scientific knowledge *A Pocket Popper* ed D Miller (London: Fontana) p 179

[13] Popper K R 1972 *Objective Knowledge* (Oxford: Clarendon) p 243

[14] Popper K R 1972 *Objective Knowledge* (Oxford: Clarendon) p 126

[15] See Miller A I 1987 *Imagery in Scientific Thought: Creating 20th-Century Physics* (Cambridge, MA: MIT Press) p 45

The 'ultimate' Popper diagram has been compiled in Lüscher E 1980 Structure in science and art *Proc. 3rd C H Boehringer Sohn Symp. (Kronberg, Taunus, 1979)* ed P Medawar (Sir) and J H Shelley (Amsterdam: Excerpta Medica) p 129 (Popper was present at this conference)

[16] Popper K R 1972 *Objective Knowledge* (Oxford: Clarendon) p 144

[17] Popper K R 1972 *Objective Knowledge* (Oxford: Clarendon) p 262–3

[18] Laudan L 1977 *Progress and its Problems* (Berkeley, CA: University of California Press)

[19] Kuhn T S 1962 *The Structure of Scientific Revolutions* (Chicago, IL: University of Chicago Press)

[20] Laudan L 1977 *Progress and its Problems* (Chicago, IL: University of California Press) p 66

[21] Laudan L 1977 *Progress and its Problems* (Chicago, IL: University of California Press) p 15

[22] Laudan L 1977 *Progress and its Problems* (Chicago, IL: University of California Press) p 48

[23] Laudan L 1977 *Progress and its Problems* (Chicago, IL: University of California Press) p 68

[24] Sarkar H 1983 *A Theory of Method* (Berkeley, CA: University of California Press) p 110

[25] Sarkar H 1983 *A Theory of Method* (Berkeley, CA: University of California Press) p 112

[26] Prior A N 1971 *Objects of Thought* (Oxford: Clarendon) p 72

[27] For an excellent historical review of 'problem-solving' see Nickles T 1988 Questioning and problems in philosophy of science *Questions and Questioning* ed M Meyer (Berlin: Walter de Gruyter) p 43

[28] Laudan L 1984 *Science and Values* (Berkeley, CA: University of California Press)

[29] Laudan L 1984 *Science and Values* (Berkeley, CA: University of California Press) p 65

[30] Laudan L 1984 *Science and Values* (Berkeley, CA: University of California Press) p 66

[31] Ajdukiewicz K 1973 *Problems and Theories of Philosophy* (Cambridge: Cambridge University Press) p 13

[32] Joachim H H 1906 *The Nature of Truth* (Oxford: Clarendon) p 73

[33] Popper K R 1972 *Objective Knowledge* (Oxford: Clarendon) p 52

[34] Watkins J 1984 *Science and Scepticism* (Princeton, NJ: Princeton University Press) p 169

[35] Grünbaum A 1976 Can a theory answer more questions than one of its rivals? *Br. J. Phil. Sci.* **27** 1–23

[36] Kuhn T S 1970 *The Structure of Scientific Revolutions* (Chicago, IL: Chicago University Press)

[37] Feyerabend P 1970 Consolations for the specialist *Criticism and the Growth of Knowledge* ed I Lakatos and A Musgrave (Cambridge: Cambridge University Press) p 219

[38] Dilworth C 1986 *Scientific Progress* (Dordrecht: Reidel) p 50

[39] Dilworth C 1986 *Scientific Progress* (Dordrecht: Reidel) p 67

[40] Simon H A 1966 Scientific discovery and psychology problem solving *Mind and Cosmos* ed R G Colodny (University of Pittsburgh Press)
[41] Baudrillard J 1983 *Simulations* (New York: Semiotext(e)) p 104
[42] Hughes R I G 1989 Bell's theorem, ideology and structural explanation *Philosophical Consequences of Quantum Theory* ed J T Cushing and E McMullin (Notre Dame, IN: University of Notre Dame Press) p 198
[43] Watkins J 1984 *Science and Scepticism* (Princeton, NJ: Princeton University Press) p 191
[44] Bachelard G 1969 *The Poetics of Space* (Boston, MA: Beacon) p xii

Chapter 5

Subjective nature of science

5.1 Cartesian view of science

The root of the Cartesian concept of nature is the subject–object dichotomy. The abstraction of the observer from the world is so closely allied to the feeling of distinction between the self and everything else that slipping into this groove is difficult to avoid [1]:

It is the subjective self who knows.

Popper makes a distinction between this kind of organismic knowledge and objective knowledge 'which consists of the logical content of our theories', which he refers to as 'World Three', Worlds One and Two being respectively the physical and mental worlds. Apart from theoretical systems, other inmates of Popper's World Three are critical arguments and problem situations [2]. Popper sees World Three as objective because although it is a human creation it has a 'domain of autonomy' [3]. Thus Popper reconciles the objectivity or autonomy of his World Three with its man-made nature [4]. In a sense he has the best of both worlds—if the reader will excuse the pun.

Poppers's problem-solving science has its objective element, which is sufficient to demarcate science from non-science by upholding an autonomy which is science's objectivity. There is a 'characteristic of people to transform everything which in truth springs from their own invention and design promptly into an objective giveness' [5]. As Mary Warnock underlines 'we cannot depopulate the world of objects by merely wishing to do so. In that sense objects are independent of us' [6].

For Latour, subjectivity and objectivity represent two extremes of a trial of strength that any view or theory has to undergo. There are subjective individuals and objective representations [7]:

Being objective means that no matter how great the efforts of the disbelievers to sever the links between you and what you speak for, the links resist.

In this scenario belief is subjective, but knowledge is objective [8]: 'Nature cannot be forced into an arbitrary set of conceptual boxes' [9].

The problem is the idea of an objective reality *out there*, with beliefs and mental states separate from objects and inhabiting an inner-realm. This essentially Cartesian language sets the stage for the problem of how to eliminate the subjective elements from science. There is of course a lot of question-begging going on here. The subject–object division itself presupposes an objective world. The alternative [10] is a theory of knowledge as construction relying instead on a knowing subject having a dialogue with external reality [11]:

The concepts that we use are not ... built into reality. Rather, concepts are human inventions.

The very notion of a scientific concept, however, has theory as the central part. It is not that there is a mental picture of a concept, like the photograph of an object, but that scientific concepts are part of a theoretical structure of ideas representing the way we look at and cope with the world.

However scientific progress is viewed, objectivity is an important issue. But since science involves scientists, the subjective element must always be present. The only way to achieve objectivity is to divorce the subjective and objective elements from each other—but this is not possible. The situation is analogous to the construction of a building. Once the structure is complete and all the builders are finished, the property stands alone and if built well will stand the test of time. Once complete, it is in the Popperian sense an autonomous object—if it falls down it is the equivalent of a discarded theory that has been falsified. Also, the impossibility of an ideal building reflects the impossibility of attaining absolute truth in a theory. But theories are not buildings, so the analogy is limited. If a theory achieves autonomy, then it has a transcendent element beyond its creators: the scientists. And this transcendent quality is very close to, if not coincident with, the truth of the theory.

The term subjective, particularly as applied to the social sciences, implies goal-directedness or intentional action [12]. The idea, in

science, is that since the universe blithely carries on seemingly impervious to our wishes, it therefore follows that science is objective. The problem is that just as in history it is the winners that write the textbooks, so in science it is the scientists who not only write the texts, but who create, interpret and develop the theories.

We seem to have come round full circle. The reason is that truth and objectivity are both absolute terms. The autonomy of a theory implies the 'death of the author' (or authors) and even though this may be a useful concept in literary criticism, in science true objectivity involves the death of the universe as well.

To continue the literary analogy, certainly a text can stand alone within its own framework and the author's life may be considered irrelevant. The reason is that the world that the text is part of and which is created by the text and the readership is based on the actual world. Even the most bizarre science fiction novel reflects its essential human construction—a good lie is one which is based on some truth.

Within the scientific context, the autonomy of a theory is in a similar sense a fiction because the world that a theory represents is based just as much on the actual world as is a fictional narrative. That is not to say that a theory is the same thing as a fictional narrative (though some may be), but that the problematic web which constitutes a theory is a result of an essentially human intellectual construct. My point is that whereas a question, which is a constituent part of a theory, is autonomous, the theory is not.

There are two ingredients which are added to questions to make theories: firstly, the answers to these questions and, secondly, the pattern of linkages thus formed. It is these elements which, because of their temporary nature, render the concept of autonomy inapplicable.

In the zetetic model, the goal is not truth but a valid combinatorial structure. And it is the logical or formal structure of this matrix which is the objective or deductive element. The questions—the vertices of the structure—are created by scientists as a result of mutual interaction with nature, and the answers—which are only tenuous links between vertices—represent relations between questions thus reflecting the scientists' theoretical model of the world. The truth value of a theory is thus a concept which is inapplicable, the deductive element applies at a higher epistemological level, i.e. the mathematics of graph theory and the combinatorial rules which relate to what constitutes a valid structure.

5.2 Translation from observations and concepts to questions

Even though scientific language often implies pursuit of a question-answering activity; this is not usually evident at the level of practical or theoretical research. A researcher will explain that he is working on a problem or looking for the solution to a puzzle. Reference may be made to a general problem—in astrophysics this may be 'the missing mass problem','the solar neutrino problem' or 'the problem of galaxy formation'. But these phrases are more in the nature of review article titles and research itself is not perceived as a genuine questioning activity. As soon as one moves away from broad generalizations to more detailed research, work consists of either the objective recording of observations or the manipulation of mathematical structures in an attempt to give the best fit to observations.

This description is, of course, far too simplistic and does not reflect the great deal of thought that goes into research. But it is very hard for a scientist sitting in front of an oscilloscope twiddling with the knobs to accept that he is actually generating questions. How can we disabuse this poor journeyman scientist of his ingrained beliefs? Simply by asking some questions: Why are you turning the dials of the machine? How will this relate to the theoretical model that you have in mind? and so on. After a few of these probing questions, any researcher would relate what he is doing and why he is doing it, and the problematic nature of his work would become apparent.

A scientist translates what he is doing into language, although for Wittgenstein this process involves a calculation or operation with words which may result in sometimes one picture, sometimes another [13]. For most scientists the fundamental unit or goal of the scientific process is the hypothesis. But a hypothesis is not just an assertion, it is also a conjecture and therefore problematic [14]. Popper sees the history of physics as a 'history of its problem situations' [15]. In the present scenario theories can be translated into networks of questions and answers, but how does this supercede the picture of theories as conflations of concepts, hypotheses and presuppositions? The problem with the latter scheme is that these three aspects of theories are not independent, but in the former, questions are autonomous units abstracted from the network, which is itself at a different level in the hierarchy.

Presuppositions of a theory need not impinge directly on whether the question–answer matrix represents a valid theory structure. Questions

Figure 5.1 Ludwig Wittgenstein. (Reproduced by permission of the Masters and Fellows of Trinity College Cambridge.)

may be black-boxed and those in the matrix, together with their accompanying presuppositions, form a structure whose validity is determined not by the adequacy or truth of the answers, but in the formal structure of the resulting graph. This enables a clearer distinction to be made between discovery and justification, the justification being purely in the combinatorial properties of the resulting network. These properties rely in turn on the fundamental idea of *relation* to which we now turn.

5.3 Relation

By considering a scientific theory as a linked system of questions and answers, we can look upon a fundamental unit of a theory as two questions linked by one answer. This unit corresponds to a relation between questions. I have discussed the formal aspects of logical relationships in Chapter 3, but here I want to concentrate on the more specialized concept of a relation between questions.

The importance of relations has waxed and waned throughout the history of philosophy since the time of Plato and Aristotle. Malebranche in his 'Search after truth' puts relations at the forefront of epistemology [16]:

The mind of man seeks out only the relations of things

and further [17]

> ... *in all questions we search only for the knowledge of some relations, be they relations between ideas or between things and their ideas'.*

The French philosopher Antoine Cournot described 'natural phenomena' as forming 'a network all the parts of which are connected with one another' [18].

With respect to the legal system, Montesquieu defined laws as [19]:

> ... *the necessary relations arising from the nature of things.*

In *First Principles*, Herbert Spencer identifies conscious thought in relational terms:

> *A characteristic of consciousness is that it is only possible in the form of a* relation.

No thought can ever express more than relations [20, 21].

At the turn of the century, the French philosopher Hamelin accepted that 'the world is a hierarchy of relations ... constituted not of things but of relations' [22], what Whitehead referred to as a Nexus or grouping of occasions [23].

Poincaré considered that the 'real' objects of nature are forever hidden from our eyes with the only reality we can obtain being [24]:

> ... *the true relations between these real objects.*

For Lacan meaning is in the signifying chain of relations between subject and object [25].

Hume's famous 'fork' divides 'all the objects of human reason or enquiry ... into two kinds, to wit, *Relations of Ideas*, and *Matters of Fact*' [26]. If an idea does not fall into one of these two categories then according to Hume it is worthless.

The importance of relations in science is highlighted by Dewey [27]:

> *Relations become the objects of inquiry since meanings are determined in the ground of their relation as meanings to one another.*

Dewey divides relations into three categories which he refers to as relation, reference and connection. *Relation* applies only to relations between symbols. *Reference* to the representation of physical parameters, and the validity of the resultant theory resides in the *connections* that exist among things [28]. Relations themselves, without regard to the character of the terms related, become the object of logical analysis.

This enables the deductive characteristics of relations to be separated from inductive inferences concerning the terms—although this separation can only be imperfect, as otherwise any random theory would be considered a valid theoretical framework simply because its graph complied with the rules of the zetetic model. The zetetic model does not provide the answers, nor does it indicate which questions should be linked with each other within the framework. This is part of the inductive process which has evolved through our reasoning faculty into the formulation of scientific theories. Any answer which links two questions can be in error and the lack of a link, which implies that two points are not related, may also be in error. Thus, the creation of an alternative theory would result in both old links being broken and new links being forged.

The non-relation of two objects is itself a relation. This apparent paradox resolves itself under closer inspection; non-relation applies when we cannot supply an answer to a question which links to another particular question. It was Russell who wrote that [29]:

> *it seems impossible to make any statement of what relation is without using the notion of relation in doing so.*

Relations have been seen as either external or internal to the terms related. Seen externally, relations are not intrinsic to objects. ‘Knowledge is perception of the connection and agreement, or disagreement and repugnancy of our ideas’ [30]. Relations are between ideas, not in them.

The internal view of relations rests on the precept that the nature of a term includes all its properties, the nature of every possible object of thought is determinedly its relations to everything other than itself [31]. The middle view is that the notion of the *internal* or *external* property of something is a matter of the degree of information conveyed, if the property in question is lacking, although there may not be agreement where there are conflicting interests and purposes [32].

To frame a question or to face a problematic situation is a necessary

Figure 5.2 Bertrand Russell. (Reproduced by permission of Mary Evans Picture Library/ Idar Kar.)

precondition of cognitive operations or inquiry 'but it is not in itself intellectual or cognitive: it is a precognitive situation' [33]. Although a scientific theory is necessarily articulated in language, it is rooted in the precognitive stage of *wonder*. I referred to this point in Chapter 2 in connection with the example of a caveman trying to come to terms with the weather.

Since *questions* are the terms being related, we may ask how they differ from other kinds of related terms. *Questions* are *ideas* rather than *things*, but they are a rather special kind of idea. A scientific question is an active process of conscious direction no matter how theoretical the entities involved. This applies also to the precognitive stage—when my cat cannot get to her food bowl because the kitchen door is shut, she copes with the situation by calling either myself or one of my family over to open the door for her. The cat, when faced with a problematic situation consciously directs her mind to her situation. There is no verbal communication and presumably no theoretical model in the cat's mind. However, there is a reduced consciousness, directed to a perceived problem. This example is more useful as an analogy than as a serious enquiry into animal psychology, but it illustrates that the zetetic model can be placed in an evolutionary setting with its roots emanating from our pre-linguistic and animal forebears. Our construction of theoretical models of the universe is our way of coping with nature.

In the zetetic model, therefore, answers fulfil the role of relations between questions. The formal or logical relationship thus established gives rise to a graph whose structure shows in a deductive sense whether the theory is validated or not. This validation is not applicable at the level of the questions and answers themselves, i.e. whether a particular answer is acceptable or not. In fact, part of the discussion and argumentation in theoretical science is often not so much at the level of the interpretation of results, but at fitting them together in a coherently valid structure. This process, which is an important part of science, is exhibited in the various ways that scientists communicate with each other, and in particular in journal publications. The problem is that the myth of the autonomy of a theory—once it is created—has the negative effect of limiting the horizons of scientists to ignore social interactions, which are every part as important in creating the network which defines a theory. Even the lone researcher on a desert island trying to put together a theory is 'plugged in' to the rest of science, if he or she wants to be taken seriously. Thus the social study of science is relevant not only for this reason but also because the way that social scientists look at this problem fits very neatly into the zetetic model.

5.4 Social studies of science

The fiction of an objective or autonomous science is manifest in the deep prejudice that some scientists have towards the social sciences. The workings of the community of researchers in a laboratory environment can be seen as a social study, much as one studies a remote Amazonian tribe, yet the clear-cut distinction between a theory and its antecedent deliberations militates against acceptance of social studies of science either as part of the whole process of theoretical research or even as part of the discussion process which leads to theories [34]. However, part of the process of science consists of interaction between scientists formally and informally in the context of their work. I draw attention to the distinction between formal and informal interactions, not because I think it is particularly clear cut or useful, but because although many accept that formal interactions (such as refereeing of papers or allocation of resources for research) might have some relevance to the whole process of science, more informal situations (such as decisions by a researcher on which problems to tackle, or the interactions between scientists working together) are not generally considered to be relevant to the actual theories produced. We have returned to the idea of autonomy of science and the context of justification and discovery division.

This situation has become untenable. Science is not an activity which is pursued in a vacuum and simply to view a theory as some transcendent text which may be used to wheedle secrets out of nature, much as a Shaman performs an incantation to cure a patient, is not realistic. Of course, theories sometimes work, but then patients often get better.

There is a mythology in science just as in any other human endeavour. For example, why should effects with a statistical significance less than three standard deviations, or approximately one chance in a thousand, be generally considered unworthy of publication? Certainly it sounds a reasonable criterion, but why three standard deviations and not 2.8 or 3.2? Because three is a nice round number! Is that the reason? Partly yes, but also a consensus has developed over a long time which, in common with most folklore, is reasonable and practical. This criterion is part of the scientific process. In viewing science we therefore have to study the social aspects in terms of the methodology of science as well as the actual content of the theories produced. The 'we' that I refer to, are of course scientists themselves. In this way, the horizons of research are enlarged and the self-critical aspect of science—which is its strength—becomes more effective.

There are a number of disparate elements which under the banner of social studies of science cohere together in a mutually productive way. Firstly, there is knowledge itself. There is the image of a lone scientist making a breakthrough—the Eureka effect. I do not mean to denigrate the importance of this type of discovery, but more often than not such discoveries are part of an ongoing consensus, which maintains the social aspect of advancement. After all, if everybody is trying to jump over the bar in a high-jump contest, then sooner or later somebody is bound to be successful. Knowledge is thus institutionalized cognition [35]. Scientists like to think of the acquisition of knowledge, particularly in theoretical science, as not being goal directed. This is a particular area which is difficult to pin down. Certainly the pursuit of knowledge and understanding is itself a type of goal, but this is not what is meant. Equally, trying to solve particular problems or explain observational or experimental results is also not the type of goal to which I am referring. We have to move higher up the hierarchy to identify more important goal-directed behaviour. I refer here to 'grand problems' which are set by and evolve within the scientific community as a whole.

Each area of theoretical physics has its agenda. What kind of cosmological model is applicable to the universe? How do we account for the masses and diverse number of fundamental particles? What kind of theory will account for the four fundamental forces of nature? Where these problems come from and why they arc important in setting the goals for current research is as much to do with the social aspects of science as the theories themselves.

I refer here to the so-called band-wagon effect. For whatever reason some problems come to the forefront at certain times and become fruitful areas of research whilst other equally valid problems fade into the background. Fashions seem to change in science just as in the clothing industry and often for reasons which are equally obscure.

The second element under the banner of social studies of science is savage or non-verbal reasoning. The animal kingdom copes with everyday life without the need for language-based knowledge, and we do not necessarily raise our own consciousness level to that of verbal reasoning for many humdrum acts. Both from the point of view of the evolution of consciousness, and the importance of precognitive features in science, the second element is part of the social equation. This aspect would normally be either totally ignored or vehemently denied as being any part of a science which aspires to expunging subjective human—let alone animal—characteristics. This denial of our evolutionary heritage is as naive as it is pointless. Naive because to ignore our own nature is to ignore 'Nature', and pointless because an understanding of precognitive features in our approach to science is only possible through examination of our own intellectual evolution.

This leads to the third element which is the evolutionary aspect of our thinking process—not individual thoughts, but the collective thoughts which make up part of the whole inductive process. There are 'basic inductive propensities ... inherent in our characteristics as organisms' [36].

Our brains may be constructed so that we induce in a particular way. Since language may have developed in a universally human fashion, so there may be a human induction to which we naturally incline. That language is central to our intellectual life is undeniable, but equally there is more to language than merely words. Language is part of our social interaction and has developed accordingly. The social dimension in science is therefore also present through the medium of language.

The zetetic model, which is a common denominator of these three elements, puts each in a slightly different perspective and in so doing emphasizes the importance of the social side of science. The network or graphical model does indeed crop up in social studies of science under the heading of scientometrics.

5.5 Scientometrics

Many scientists tend to regard the social study of science as not germane to real problems because it is not directed at the scientific content of a theory but at matters considered peripheral. Bibliometric analysis is seen as a tool for politicians, science policy makers and funding institutions, rather than as a direct aid to scientists. In the present context, since a divide is created between the content and structure of scientific theories, on the structural side an analysis of relationships looks promising as a useful tool for theory analysis.

Scientometrics is concerned with measurements of the outcomes of the science process. A particular area of interest in relation to the present work is that of bibliometric analysis of scientific texts [37]—published papers, books, conference proceedings. They are the main channels of communication in science, and so analyses of these texts provide information on the communication patterns both within science and beyond to areas such as the business community and government, which for science are transepistemic [38]:

> *There is no outside of science but there are long, narrow networks that make possible the circulation of scientific facts* [39].

Scientometrics sees the scientist as an entrepreneur enmeshed within a network of constructed facts. The history of the changing pattern of the associations which form the actor-network, provides the key to the analysis of the dynamic of science. This actor-network is the analogue of the zetetic model but at a different level in the hierarchy. Whereas a scientific theory is constructed as a network of questions and answers as a result of a theoretician interacting with nature, the resultant communication, criticism and possible acceptance of the theory within the community at large is reflected in the actor-network of which the theoretician is an integral part. The ultimate verbalization of the precognitive sense of wonder is a level of the hierarchy that determines the theory and enables it to become an integral part of the intellectual fabric called science. At one level in the hierarchy, ideas, discussion and all writings which are both behind and within the theory are encompassed as part of the fabric of science.

The network of different scientific texts is mapped by bibliometric analysis just at the level where these texts compete with each other in forming the scientific consensus. In the market-place of science the actor-network entrepreneurs sell their goods. Some achieve popularity and some fade into obscurity—some associations are widely adopted, and some grow weaker. It is by counting associations amongst a large number of texts that the solidity of this actor world and its linkages are measured, and changing patterns also emerge as these linkages change over time.

One method used to map these networks is that of co-word analysis [34]. There are three stages which are needed to reveal the salient features of a network. Firstly, individual texts are reduced to a string of key or signal words with the aid of abstraction and documentation services. Secondly, a file of articles is established using an appropriate set of criteria, and a lexicon is formed of all keywords within the set. The number of documents which contain particular keywords gives a measure of the relative importance of each of the words. Furthermore, the number of documents which contain pairs of keywords represents the relative strength of the correspondence between keywords. This leads to the construction of a co-occurrence matrix which contains information about the aggregated network of problematizations. Keywords which co-occur significantly may be connected to each other in a graphical representation. By studying the patterns of co-occurrence, features such as black-boxing are revealed in an interesting and informative way.

There are, of course, problems in that the bibliographic analysis itself introduces a filtering element on the original texts by abstracting keywords. However, provided scientific texts do not employ highly metaphorical language, which is usually the case in practice, then distortion should not be significant.

Another simplifying procedure is to set thresholds on the occurrence and co-occurrence of words. The effect of this is to focus attention on the more significant effects, without submerging the picture in a sea of words which are little more than passing references. Of course, repeated analyses over periods of time will show a changing picture, with some keywords dropping out and others rising in significance and thereby entering the system.

By joining together keywords that significantly co-occur, a directed graph can be drawn, so that if a keyword A occurs more frequently than B, but A and B occur together a significant number of times, then the link B $\rightarrow$ A is established.

There are of course no specific scientific questions as such that scientometric analysis addresses, however, the important word markers which identify the changing pattern of the science process are areas of focus of problems which are current. It is not much of a translation to see that what in one context is a keyword marker to a problem area, in another context, is just a question, though not explicitly defined. The point about the way bibliometric analysis has developed is that it displays what, according to the present model, is a fundamental problem- or question-centred pattern of science.

Apart from the value of such bibliometric methods of research in the social sciences, the problematization-network basis on which the work is grounded mirrors scientists' approach to their world and the theories about it which they create.

Viewed in terms of the zetetic model, scientometric analysis thus acquires a theoretical model which underlies it. The way in which theories naturally are amenable to this kind of bibliometric analysis becomes a part of the way science itself develops.

The question-generating activity shows at a deep level the way in which science is grounded in ourselves both in the way our brains have developed and in the way that science becomes a particular aspect of the much more general phenomenon of human thought.

Instead of being regarded as adjunct to or peripheral to science, the social dimension reveals itself as internal to the whole issue and an important element which cannot be objectified away. As the emperor Marcus Aurelius asserts [40]:

> *I choose to do what is according to the nature of the rational and social animal. The intelligence of the universe is social.*

The social aspects of science are thus part of the whole process of science which in turn, as a human enterprise, has evolved within us as part of our human nature. I explore this point more fully in the next chapter.

Notes and references

[1] Popper K R 1972 *Objective Knowledge* (Oxford: Clarendon) p 73
[2] Popper K R 1972 *Objective Knowledge* (Oxford: Clarendon) p 107
[3] Popper K R 1972 *Objective Knowledge* (Oxford: Clarendon) p 118
[4] Popper K R 1972 *Objective Knowledge* (Oxford: Clarendon) p 159
[5] Hübner K 1985 *Critique of Scientific Reason* (Engl. Transl. J R Dixon and H M Dixon) (Chicago, IL: University of Chicago Press) p 23
[6] Warnock M 1970 *Existentialism* (Oxford: Oxford University Press)
[7] Latour B 1987 *Science in Action* (Milton Keynes: Open University Press) p 78
[8] Latour B 1987 *Science in Action* (Milton Keynes: Open University Press) p 182
[9] Kuhn T S 1970 Reflections on my critics *Criticism and the Growth of Knowledge* ed I Lakatos and A Musgrove (Cambridge: Cambridge University Press) p 263 and footnote
[10] Arbib M A and Hesse M B 1986 *The Construction of Reality* (Cambridge: Cambridge University Press) p 171
[11] Brown H I 1988 *Rationality* (London: Routledge) p 180
[12] Nagel E 1982 *The Structure of Science* (London: Routledge and Kegan Paul) p 473
[13] Wittgenstein L 1958 *Philosophical Investigations* (Oxford: Blackwell) section 449
[14] Meyer M 1980 Science as a questioning process *Rev. Int. Phil.* **34** 49–89
[15] Popper K R 1982 *Quantum Theory and the Schism in Physics* (London: Hutchinson) p 160
[16] Malebranche N 1980 *The Search after Truth* (Engl. Transl. T M Lennon and P J Olscamp) (Columbus, OH: Ohio State University Press) p 252
[17] Malebranche N 1980 *The Search after Truth* (Engl. Transl. T M Lennon and P J Olscamp) (Columbus, OH: Ohio State University Press) p 489
[18] Cournot A A 1956 *An Essay on the Foundations of our Knowledge* (Engl. Transl. M H Moore) (New York: Liberal Arts) section 67, p 97
[19] Montesquieu Baron De (Charles De Secondat) 1962 *The Spirit of Laws* (New York: Hafner) p 1, also see introduction p xxxix
[20] Spencer H 1946 *First Principles* (London: Watts Thinkers Library) p 62
[21] Spencer H 1946 *First Principles* (London: Watts Thinkers Library) p 70
[22] Hamelin O 1907 *Essai sur les Eléments Principaus de la Représentation* (Paris: Félix Algan) (quoted in Copleston F 1985 *A History of Philosophy* (New York: Doubleday) vol IX p 148)

[23] Whitehead A N 1942 *Adventures of Ideas* (Baltimore, MD: Penguin) p 195
[24] Poincaré H *Science and Hypothesis* (New York: Dover) p 160
[25] See Richardson W J 1988 Lacan *Continental Philosophy I* ed H J Silverman (London: Routledge) p 123
[26] Hume D 1936 *Enquiries* ed L A Selby Bigge (Oxford: Oxford University Press) IV. I.20 p 25 (originally written 1777)
[27] Dewey J 1939 *Logic: The Theory of Inquiry* (London: Allen and Unwin) p 116
[28] Dewey J 1939 *Logic: The Theory of Inquiry* (London: Allen and Unwin) p 55
[29] Russell B quoted in Adler M J and Gorman W 1982 *The Great Ideas II* (Chicago: Encyclopedia Britannica) p 569
[30] Locke J *An Essay Concerning Human Understanding* (London: Allen and Unwin) vol IV, I.2
[31] Blanshard B 1939 *The Nature of Thought* (London: Allen and Unwin) vol II p 452
[32] A comprehensive discussion of the internal/external relations debate appears in an article entitled 'Relations, internal and external' Rorty R M 1967 *The Encyclopedia of Philosophy* ed P Edwards (New York: Macmillan)
[33] Dewey J 1939 *Logic: The Theory of Inquiry* (London: Allen and Unwin) p 35
[34] Callon M, Law J and Rip A 1984 *Mapping the Dynamics of Science and Technology* (New York: Macmillan) p 108
[35] Barnes B 1983 Knowledge and cognition *Science Observed* ed K D Knorr-Cetina and M Mulkay (London: Sage) p 43
[36] Barnes B 1983 Knowledge and cognition *Science Observed* ed K D Knorr-Cetina and M Mulkay (London: Sage) p 38
[37] Chubin D E and Restino S 1983 Research programmes and science policy *Science Observed* ed K D Cetina and M Mulkay (London: Sage) p 57
[38] Knorr-Cetina K D *Science Observed* ed K D Cetina and M Mulkay (London: Sage) p 132
[39] Latour B *Science Observed* ed K D Cetina and M Mulkay (London: Sage) p 167
[40] Aurelius Antoninus M 1902 *Thoughts* (Engl. Transl. G Long) (London: Bell) p 68

Chapter 6

Evolution and intelligence

6.1 Evolution of questioning ability

The most important biological characteristic of our species is our use of language. This non- or extra-biological method of passing information down through the generations has led to an extraordinary developmental explosion. But language is more important than simply being a means of communication—all animals communicate. Our particular ability lies in the questioning activity that underlies our language: 'We can pose abstract and useless questions' [1]. There is certainly a non-verbal questioning process in other animal species, but question generation has become central to human thought processes. Even in the non-verbal areas of intellectual function such as art and music, the questioning force may be present. I am not saying that a painting literally asks a question, but in order to have an effect on people a great work of art may provoke a subjective response and this may engender a sense of puzzlement or wonder. This problematic dimension thus pervades both the linguistic and non-linguistic realms.

In the present work, scientific theorizing is seen in the context of a hierarchy of questions whose matrix formulation obeys certain rules which give general validity to the structure. In a different context, language itself is seen as a tree-like hierarchy and this feature, common to all languages, may be part of our genetic make up. Such a universal grammar presupposes an innate component of the brain which is a genetically determined language faculty [2], although one can accept a tree-like hierarchical structure common to all languages that has developed into an increasingly complex system in a bootstrap fashion without the need for specific genetic hardware [3]. Whichever type of evolutionary context is perceived, there is a basic tree-like hierarchical structure to language. The structure of language can be compared with that of knowledge. The growth of knowledge according to Popper is a Darwinian natural selection of hypotheses: 'From the amoeba to Einstein ... we try to solve our problems' [4]. This echoes Russell's observation on mankind's origins [5]:

> *Man has developed out of the animals, and there is no serious gap between him and the amoeba*

and further

> *... there is much in the analysis of mind which is more easily discovered by the study of animals than by the observation of human beings.*

We understand the world through problems and our tentative solutions are the web of questions and answers which we create. Generative hierarchies, such as in language or knowledge, are biologically grounded. The constantly repeating pattern of nested hierarchies ultimately includes itself and perhaps gives rise to consciousness. The matrix of knowledge is essentially a human construct and marks what it is to be human. We think the way we do, because that's the way we are!

One may ask the sceptical question: how could we describe any other way of thinking other than in human terms? This is partly justified, but we can describe the thought processes of other species. Furthermore, we can discuss the type of logic underlying digital computers. I am not saying that other species, or digital computers, are able to conduct rational discourse. However, just because the wheel may be the most efficient means of vehicular locomotion, it does not follow that evolution must endow our species with that particular ambulatory means. There may very well be ways of rationally coping with nature other than thinking in terms of hierarchical networks. It is just that this is the way our species has developed, there is no reason to believe that evidence for this development is not present in other species.

6.2 Computer simulation and artificial intelligence

One effect of increasing computing power is the proliferation of computer simulation studies as a research tool in the physical sciences. Such studies generate and analyse what may be termed 'non-experimental' data.

These data have been widely used in many fields outside of the physical sciences, such as artificial intelligence (AI), economics and sociology, where individuals and whole societies have been modelled by computer. These simulations generate consumers and producers living their lives, working, voting and doing all the things ordinary people do in the real world.

It has been recognized since the work of Dreyfus and others [6] that there are problems in a straightforward application of computers to these types of areas: problems which are to do with a fundamental difference in the way in which a computer makes calculations compared to the real-world behaviour of people and societies.

What I wish to highlight is that many of these and other related problems also apply to computer simulation studies of systems in the physical sciences.

There are a number of different kinds of simulations and the same principles cannot be applied to them all. However, one characteristic that they all share is the use of random or, more accurately, pseudo-random numbers. Random numbers may be required either as initial input or during a computation. This is representative of our lack of knowledge in a particular area which we feel is justified by the substitution of such 'generated' data.

The first such computer analysis, dubbed 'Monte Carlo', was used to solve mathematical problems whose exact solution appeared inaccessible. Monte Carlo analysis originated in statistical sampling techniques which were too impractical to be widely used before modern computers. Stanislaw Ulam and John von Neumann were the first to formulate a Monte Carlo computation by electronic computer—the name Monte Carlo was first coined by N Metropolis [7].

In these cases, processes in the world are mimicked by the running of a computer program which incorporates random numbers [8]. This mimicry can be tremendously effective and compelling, creating a whole environment—cyberspace—clothing the real world [9]. The 'mathematization' of nature in this way dates from ancient times, but in its modern sense nature becomes a 'mathematical manifold' [10]. The software takes on the twin roles of 'model' and 'diorama' [11]—a small-scale model of a theatrical setting.

In terms of the simulation of human actions and behaviour, virtual reality becomes theatre [12]. It might seem rather a large jump from solving equations by computer to the theatre but such is the wide net of computer applications that Shakespeare's aphorism 'All the world's

Figure 6.1 Stanislaw Ulam. (Reproduced by permission of AIP Emilio Segrè Visual Archives.)

a stage' takes on a new meaning within AI.

The theatrical metaphor is quite apposite, in particular with regard to the element of contrivance in a theatrical production and the false sense of the passage of time and causal sequences, which also can be a feature of a computer simulation.

There are a number of problems associated with the use of computer simulation as a research tool [1]. The data represented in a computer are discrete and deterministic and it is an 'ontological' assumption that the real world can be represented in this way. This difference between the analogue 'real' world and the digital 'virtual' world is also highlighted by the 'grain problem' [13], where objects in the real world are 'simplified' by representation as discrete data. This problem is just as important in simulations of physical processes as in AI.

A particular calculation which often features in simulation work is the n-body problem [14]. This consists of bodies moving under the influence of their mutual gravitational attraction, whose evolution in time is modelled by computer.

Figure 6.2 John von Neumann. (Reproduced by permission of AIP Emilio Segrè Visual Archives.)

In carrying out a computer simulation of an n-body problem the calculation has to proceed along a finite number of time steps. There is always a compromise between increasing the number of particles involved in the calculation or decreasing the time interval between consecutive steps.

These problems are well recognized and their effects have been investigated, but there are further problems which are more insidious and less open to straightforward analysis.

Quite apart from digital computers, simulation is something that we ourselves indulge in—and to an extent so does the rest of the animal world. We try to anticipate the future nearly all the time, as reflected in our everyday actions and when we plan, make decisions, vote and dream [15]. In more abstract terms we share with the rest of the animal world anticipation of the future by virtue of the fact that we are all complex adaptive systems [16]. It is this notion of control and feedback that also lies at the heart of simulation by the use of digital

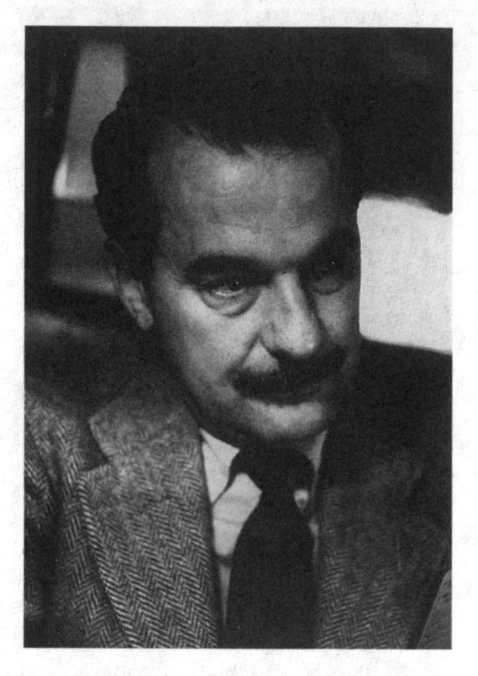

Figure 6.3 Nicholas Constantine Metropolis. (Reproduced by permission of AIP Emilio Segrè Visual Archives.)

computers [17].

AI computer simulations display this element of control and feedback. Many simulations are of physical systems which proceed sequentially, with each part of the process strictly regulated by the historical development up to that point in time. However, computer simulations just do not work that way. When an author presents a simulation in a publication, there have invariably been a large number of simulations undertaken, the results of which have been used to modify, augment and justify the work. A computer simulation, therefore, should always be seen as a selected representative of a cluster of simulations. The consequence of this is that a subtle but powerful form of feedback is involved. In the case of a physical process, as described, there can be no feedback or control in the real world as this would violate causality. Whereas in the 'virtual' world we can keep going back and re-trying simulations, in the 'real' world you only get one shot!

We utilize our own 'simulations' to make the probability of certain future outcomes more likely—this is part of the struggle for survival. We do not, of course, violate causality, although when a particular 'planned' outcome actually occurs, we have achieved a pseudo-violation of causality.

In the purely physical processes simulated by n-body calculations, there is neither goal-directedness nor pseudo-violation of causality nor any kind of feedback mechanism at work, characteristics which are not shared by a computer simulation. There is therefore a tension between the goal of a researcher to simulate a particular physical process in nature and the problems caused by introducing a non-physical feedback mechanism.

The effects of this are to introduce filtering mechanisms which select appropriate model simulations. For example, a simulation may produce results which numerically diverge to infinity or are so unstable that no meaningful result emerges. But this situation is extremely rare because models can almost always be 'tweaked' so as to result in a well-defined answer.

This highlights another problem for simulation work in that there is invariably some answer available in any simulation calculation. This contrasts with theoretical research in general where most of the time the researcher does not know whether the line that he or she is following will yield a result at the end. Publications represent those lines of research out of which some sort of result has emerged.

In the case of research involving simulation calculations, there is a built-in guarantee that a well-defined result will emerge. This encourages a flood of simulation publications which are not subject to the usual 'survival of the fittest' restraint. A determined modeller with a computer and a half-baked theory with plenty of parameters that can be 'tweaked' can produce a simulation which fits tolerably almost any set of physical data. The resulting paper with a plethora of diagrams eye-catchingly drawn by computer may not be eliminated by a less alert referee.

This is an example of a filter mechanism not working when it should, but the opposite effect also occurs. In this particular situation the feedback caused by performing and selecting simulations introduces a non-physical element. In particular, a published simulation, though perfectly correct, may represent such a small class of simulations within the whole that the probability of it being physically applicable may be vanishingly small.

In other applications the feedback loop can be important and useful.

In the arts there is a feedback between the audience and the playwright and between the reader and the author [18]. This dialectical dimension is as fundamental to the arts as simulation. However, in the case of physical processes, such feedback in the simulations can introduce traps for the unwary.

In the jargon of AI, nature is crashproof, whereas programs are not. A program can halt its operation before completion for a number of reasons. Apart from an error or 'bug' in the program, the generation of numbers which are too large can initiate shut down and termination of a program during execution. In particular, in simulations where pseudo-random numbers provide an important input, a particular and unforseen chance combination of inputs may result in the generation of excessively large (or small) numbers and lead to a crash.

This may happen very rarely. In fact it may be so rare that a simulation may be run many times without it happening at all. The problem here is that one never knows when such a crash might occur. One could say that this problem can be ignored and those simulations which actually crash are simply discarded. But this introduces an element of 'selection' which is not present in nature.

The notion of a program crash has a certain parallel with Leibniz's principle of continuity [19]:

Nature makes no leaps.

Leibniz (figure 6.4) meant that in nature there are no real discontinuities or sudden changes—in the present context nature's 'program' is uncrashable. This principle would have to exclude the start and end of the universe—the 'Big Bang' and 'Big Crunch'.

One aspect of n-body and many other physical systems is their chaotic structure. I refer to the sensitivity of such systems to initial inputs—poetically referred to as the butterfly effect. Chaos brings its own special problems in simulation work. Firstly, there is the good news: in a chaotic system the range of inputs often, in a statistical sense, does not change the outcome. More accurately, the range of outputs follows a statistically well-defined distribution from which useful results emerge.

Secondly, there is the bad news: for many systems this is not the case. The news, following my metaphor, gets even worse: many systems appear to belong to the 'good news' class in that output parameters seem to follow some reasonable distribution, but occasionally certain input parameters have radical effects. Normal

Figure 6.4 Gottfried Wilhelm Leibniz. (Reproduced by permission of Mary Evans Picture Library.)

interpolative functions are unreliable in these cases and, furthermore, it is not always apparent which systems are going to 'misbehave' in this way.

This effect does not necessarily manifest itself by way of input parameters just having special values. There may be subtle correlations between a number of input parameters which might push the whole system into a novel and radically different state. The correlations may be unavoidable, unforseen and even just due to chance values assigned by the random-number generator used in the simulation.

All systems have to start somewhere. In simulation work this corresponds to the setting of initial conditions. Another problem in chaotic systems is how a system can achieve a state where all the 'input' parameters have the values they do. This entails working a simulation backwards in time—where all of the above problems equally apply. There is an old adage in computer simulation work that you cannot really calculate anything without first knowing the answer!

Simulation studies have gone beyond merely empirical or numerical experiments but are 'thought experiments' [20]. The 'what-if' situation when successfully applied imparts a heightened reality to the generated data. The data are more than just numbers—you can see the created objects on the computer screen. I do not here just refer to the 'virtual' kind of reality whose sensually interactive nature sells computer games. There is the more subtle putative reality of the data which, though it may be instructively and usefully employed, remains essentially non-experimental, non-observational and most importantly non-physical.

An image of a cluster of galaxies may be a genuine picture or a computer-generated simulation. Just because you cannot tell the difference does not mean that it is not important. If you take the computer program as your theory [21], then a complex enough program can always produce data that mimics well enough that produced by a restricted set of observations [22]. Since you cannot observe an infinite number of data, all measurements belong to a restricted set and the avoidance of this trap is what science is all about. In simulation work this trap is all the easier to fall into.

There is a fundamental paradox in that the (quite deliberate) way that 'virtual' data are portrayed so as to appear indistinguishable from 'actual' data is in opposition to the best scientific model-building methodology which should go beyond appearances. Seeing is definitely not believing!

6.3 Neural nets

I have deliberately avoided any teleological implications in discussing the evolutionary context. Although coping with nature may be seen as a goal, in an evolutionary context it is no more than an animal's response to the sieve of natural selection. The goal of any AI machine is to replicate, or even exceed in capability, the rational processes which we associate with the human mind. How far AI has succeeded is outside the scope of the present work, but there are certain aspects of AI which merit attention, particularly neural networks, which are homologically similar to the zetetic model.

Neural nets have evolved out of connectionism which involves building computational devices by directly modelling neurones. The basic units of a neural net can be thought of as the vertices of a graph and the strength and number of all the connections describe the system. Since the strengths of the interconnections can be altered, neural networks can learn and knowledge is stored in a distributional way. Neural networks belong to the general class of knowledge representation formalisms also referred to as semantic networks which are graphs where the arcs represent relations between vertices which represent objects, concepts or situations [23].

The difficulty with these structures is that they do not address the fundamental question of how to recognize a problem in the first place [24]. It is necessary to find and formalize the problem before one can attempt to solve it. Many AI machines have the problem to be solved incorporated in the hardware or software *ab initio*. This circumvents the initial problem recognition, but in so doing abandons that aspect of intelligence which I may loosely refer to as curiosity [25]:

> *The ability to wonder why, to generate a good question about what is going on . . . is the heart of intelligence.*

As Dewey puts it: 'Curiosity set free is discovery systematized' [26].

6.4 Curiosity

There are various stages of curiosity in man and in the animal kingdom. Dewey identifies a non-intellectual curiosity which at the lower level is exhibited by animals in their abundant organic energy ceaselessly reaching, poking, pounding and prying. At a higher social level a young child will continually ask 'Why?' and 'What is . . . ?' questions, although behind these questions is a naive curiosity with evidence only of a very reduced consciousness of rational connections. The highest stage of intellectual curiosity reflects an interest in problems and is entirely language orientated [27].

However one views the different levels of curiosity, there is certainly a wealth of characteristics in humans and animals, right down to the humble amoeba and beyond, which can loosely be identified under the general heading of curiosity. This spectrum reflects an evolutionary sequence of a characteristic just as identifiable as for example an appendage (arm, leg or tentacle). The problem at present is that there are no theories of brain function which are far enough away from being just 'metaphors in search of a genuine theoretical articulation' [28]. One cannot therefore tie up very easily brain form with function. Although, if we look away from the specific example of curiosity to the more general case of hierarchical behaviour, I think that there is a strong case in linking this with the hierarchical function and form of the brain.

In the zetetic model the rules which govern the creation of valid structures do not put the semantics into the system. It is the problems or questions created which provide the meaningfulness resulting from the minds intentionality towards nature: this intentionality being directed both towards and from within the world. In recognizing problematic situations, scientists weave a net which embodies physical theories. AI traditionally started too far down the hierarchy—by concentrating on problem solving and not problem recognition, ignoring a fundamental aspect of intelligence.

It is hoped that by modelling AI machines on the workings of the human brain, they will naturally aspire to meaning and consciousness. For instance, neural nets share properties with the human brain: they have a learning ability, are superior to other computers in pattern recognition, and limited malfunctions do not necessarily cause the system to crash [29]. The neural net consists of vertices with several inputs and outputs. When a weighted sum of the inputs exceeds a specific threshold the outputs are triggered. The net is characterized by the way the units are connected and how these connections are changed through the learning process [30].

Goldman [31] lists varied cognitive functions which can all be represented in a hierarchical structure. Apart from those based on language (problem solving, grammar, meaning) he includes examples from visual perception, music and animal behaviour.

Our brains are therefore structured both physically and functionally as hierarchical systems, and this characteristic continues in a reduced sense throughout the animal kingdom. The zetetic model represents a particular aspect of the way we try to understand nature and as a model is itself part of the hierarchy that determines our mental lives.

There is a peculiar kind of self-reference involved with consciousness. We have the ability to reflect on ourselves within the world. In an abstract sense this is also a characteristic of a hierarchical system. If I represent what I am saying about the world via physical theories which are in the form of a semantic net, I can reflect on the fact that a semantic net is a hierarchical structure. That latter reflection is itself part of the hierarchy which thus includes itself. The way that all these thoughts are organized in my brain and the physical structure of my brain as a network of neurones facilitates hierarchical reasoning and the result is a blurring of function and form. An important feature of a hierarchical structure is that if you add a tier to the hierarchy it retains its hierarchical form. It is this self-referential characteristic common to hierarchies, brains and minds which inextricably ties in hierarchies as the putative mathematical model for rationality.

The question being addressed is how and why science developed. I do not mean just the rather narrow but important pursuit of modern science in the last three hundred years or so, but the more general human approach to the world which has come under this heading since time immemorial. In one sense the question could be taken as pointless, since it implies that something other than science could have developed, in which case one could still be asking the same question about a unique event. On the other hand, I believe that one can put forward a meaningful reply to this question by looking at the roots of science as a human endeavour further rooted with other animals in a common evolutionary past. Just as the drives of hunger and sex are behind the necessity to survive and reproduce, so curiosity is linked in a similar way with science.

We can of course continue this process and ask: Why curiosity? At this point we are caught up in a double problem. Firstly, curiosity implies a directedness by the subject towards something else, rather than behaviour which simply resembles curiosity. A distinction thus has to be made between genuine curiosity and trawling. Secondly, and leading on from this, there is a danger of anthropomorphism where we read into animal behaviour motives which are in our own minds rather than in the animal's. Treading a careful path to avoid falling into this double trap, I would venture to say that curiosity is basic to survival and stands behind even the drives of sex and hunger. I mean this in the sense that at its most basic level of 'go and search', curiosity is the means whereby the other drives can find fulfilment. We see in the jerky random movements of a baby the first tentative interaction with the world and even though the feedback is at first poorly maintained, this trawling of stimuli is the precursor to curiosity.

We may try to look behind scientific activity to see what lies beneath, perhaps to seek firmer ground. As Chalmers points out [32]:

> *The problem of defining some unchanging essence lying behind the social, cultural and historical differences (between people) is notoriously difficult. It is no doubt an essential feature of humans that they are able to think and sense. However, it is not likely to be fruitful to seek the nature of science in whatever is universal in those capacities for the simple reason that, whatever the enduring*

capacities of humans might be, the reasoning, observational and experimental processes involved in science change and evolve historically.

But we can step back a few paces to other levels of the hierarchy and look at the underlying human motivational background to intellectual or creative activity. According to Koestler [33]:

All creative activity ... (has) a basic pattern in common: the co-agitation or shaking together of already existing but previously separate areas of knowledge, frames of perception or universes of discourse.

So are we all basically movers and shakers and have we in a sense just come round in a full circle back to Malebranche's brain traces with their mutual linkages, referred to in Chapter 2? Perhaps, but we have to keep peeling away at the onion skin. Behind intellectual activity is curiosity.

The problem with looking at curiosity, wonder, desire [34], exploratory instinct or any other description is that it is so much part of our lives and so transparent an activity that it tends to be passed over.

We think of wonder as a state of mind which for Socrates 'is the feeling of a philosopher, philosophy begins in wonder' [35]. Curiosity or interest in novelty is behind this state of mind and defines a *hypothetical construct* [36] whose observed behaviour is termed exploration. Berlyne succinctly puts it by asking [37]:

How (do) animals behave ... when they have nothing particular to do (?).

This underlies the problem of trying to explain something which appears to be fundamentally acausal. Descartes thought of animals as automatons, inert unless disturbed, with motivation being supplied by hypothetical forces designated drives [38]; we now see animal and human behaviour in a different light.

One must be very careful in discussing behaviour in humans and the rest of the animal kingdom. Firstly, in broad terms it is difficult to be precise in delineating exactly where *human* ends and *animal* (i.e. non-human) begins with regard to behaviour, as opposed to who is doing the behaving. This is particularly problematic since we are considering things from a human point of view. Secondly, when concentrating on specific behaviour characteristics, we have the problem that similarly observed behaviour may arise out of completely different underlying motivations. This elitism on the one hand and anthropomorphization on the other are almost impossible to exorcise.

The most serious problem, however, is an extreme vagueness about what it is exactly that we are talking about. Let me try to make the position clearer. We observe in ourselves and other animals exploratory behaviour for which there appears to be no immediate motivational reason. We describe this urge in ourselves by the term curiosity, and we refer to states of wonder or possibly astonishment and surprise. But what is it that is behind these outward appearances and descriptions? This is the point where things become vague.

I can underline this problem in another way. One of the differences between computers—at least the ones we have at the moment—and animals, is that if you switch on a computer but do not run any software, then the machine will just sit there doing nothing indefinitely. On the other hand, if you satisfy all the needs of an animal—water, food, sex, temperature control, etc—then the animal will not just sit there doing nothing all the time, sometimes it will explore.

Most of the studies of animal behaviour involve laboratory animals running through mazes and I shall refer to Toates' book on animal behaviour [39]. Exploration is often related to other more definable strategies, for example a search for food. A hungry rat will leave a familiar home cage for a new location more quickly than a food-sated rat [40]. But it is clear that when every conceivable motivation is missing, there is still an important intrinsic motivation which operates [41]. This, however, can be a very powerful urge. Toates cites Darwin who noticed that chimpanzees despite exhibiting obvious terror when shown a snake in a box, nonetheless occasionally peeped inside the box when it was placed inside their cage [42].

Curiosity in humans is a much more complicated affair. In spite of our being able to communicate, this does not help in answering the question of how curiosity is grounded. By studying behaviour in children we can try to bridge the gap between humans and other animals and at the same time try to understand intellectual or cognitive curiosity or what Peirce refers to as gnostic instinct [43].

Even at birth a baby has short periods of spontaneous activity [44]. This is obviously to do with exercise and development of muscle control but it is also an early attempt to explore, which later on is extended to play. As a child matures, attempts to master the physical environment are more important with an experience of pleasure in achievement and in personal competence [45]. In older children social relations become more important, but in a more basic sense the reliance on a mother and other people for other social relations develops from birth [46]. The adult world is much more complex and coping with the world both in a physical sense of satisfying everyday needs and in a social sense—relations with others—brings forth a complex interplay of behavioural motivation.

But curiosity is much more than merely what is left over after identifiable causes are removed. It is the result of a more basic activation exhibited throughout the animal kingdom right down to the amoeba. Seen in this way, this activity is a prerequisite not just an addition to other behaviour. Out of this basic activation or seeking-mechanism all the exploratory and ultimately intellectual curiosity derives.

This explanation is beginning to sound a bit like a drive theory, but it is quite different because of the level at which it applies. Drives are seen as an innate need which an organism has to satisfy. Thus the thirst drive is satisfied by drinking. I do not see curiosity as simply a drive which has to be satisfied by the organism embarking on exploratory behaviour. Rather it is an activation present from our earliest cellular origins which has evolved throughout the animal kingdom and which in ourselves we refer to as curiosity.

6.5 Structuralism

The essence of the zetetic model is the identification and abstraction of structure which underlies science as a process. In this sense there are certain parallels with the general movement of 'structuralism'.

Generally structuralist ideas apply to literature and language, and apart from child development (Piaget) and anthropology (Lévi-Strauss) do not seem to have been applied directly to the sciences [47]. In fact, structuralism has been seen more in terms of the divisions or opposition between science and literature rather than as a process applied to both in common [48], the structuralist parts being applied to the 'language' of science.

The various movements against and beyond structuralism 'unmasking' false truths and discrediting static models of human existence [49] have largely passed science by as a result. But the objectification of nature is itself a fiction and the separation between science and literature is more applicable as a cultural rather than as an intellectual process,

Figure 6.5 Claude Lévi-Strauss. (Reproduced by permission of Jaques Hailliot/Camera Press.)

culture being according to Barthes that '*bizarre toy that History never breaks*' [50].

In Chapter 10 I further discuss the question of science and literature but I turn here to those structuralist areas of the 'soft' sciences, namely anthropology and Piaget's work on the development of intelligence in children.

6.5.1 Structural anthropology

I take as the heading for this section the seminal work by Claude Lévi-Strauss (figure 6.5) on human societies [51]. He sees common logical structures of thought to all cultures which lie behind the enormous cultural and social diversities that are observed among peoples. Lévi-Strauss's framework is that of the intellectual unity of mankind.

An insight into the relationship between societies and the world in which they live is given by the analysis of myths or more precisely the abstraction of mythical structures. To Lévi-Strauss myths were an example of 'untamed' thinking and his analysis of various myths showed not only an elaborate internal structure but also the relationship of different myths with each other.

Seen in this way, myths should be analysed in groups rather than individually, with systematic differences as well as similarities between myths shedding light on deeper structure.

Another element of the analysis of different cultures is that of kinship or marriage rules. Lévi-Strauss's codification and classification of an enormously complex mass of data on customs and practices went much further and was more successful than anything that had previously been attempted.

6.5.2 Piaget's structuralism

In studying the development of intelligence in children, Piaget (figure 6.6) sought to shed light on the underlying processes of knowledge acquisition both in terms of human intelligence and in more general abstract terms of biological structures.

For Piaget, structures are systems governed by laws. Structures are not static but form as a result of transformations which yield dynamic structures. These are not put together element by element but are conceived as a dynamic whole which has arisen from a series of transformations. To a first approximation Piaget defines this system of transformations as a structure [52].

The final element that Piaget adds to this edifice is that of self-regulation or autonomy where there is a tendency for equilibrium, so that each transformation of a structure is orderly in the sense that it results in a new equilibrium. This process of feedback or 'equilibration', as Piaget calls it, results in the acquisition of knowledge by an organism [53].

In his system the 'invariant' quantities are not cognitive structures, which change over time, but the functional processes of adaption and organization which are the same throughout all developmental stages [54].

In common with other structuralist thinkers humans are seen as constructive organisms—Piaget defines knowledge as 'essentially construction' [55].

Piaget has his critics [56], though the link provided by Piaget's structuralism between AI, cognitive psychology [57] and philosophy of science [58] has been extremely fruitful.

Figure 6.6 Jean Piaget. (Reproduced by permission of R Crane/Camera Press.)

The main point of this chapter has been that the zetetic model, which has been introduced as a rule-governed hierarchical structure for theoretical science, is part of a much wider and more general biological structure. As far as the model itself is concerned, it firmly grounds it within the evolutionary process of our brains and epistemology. As for science, this is no longer an autonomous area of intellectual pursuit, but is a certain aspect of the life of the mind in which all share to a lesser or greater extent.

In the next two chapters I put the theory of the zetetic model into practice by considering a number of examples of theories and how the present model applies to them.

Notes and references

[1] Gregory R L 1981 *Mind in Science* (Baltimore, MD: Penguin) p 179

[2] Chomsky N 1986 *Knowledge of Language* (New York: Praeger) p 3
[3] Tennant N *Minds, Machines and Evolution* (Cambridge: Cambridge University Press) ed C Hookway p 87
[4] Popper K R 1972 *Objective Knowledge* (Oxford: Clarendon) p 261
[5] Russell B 1921 *The Analysis of Mind* (London: Allen and Unwin) p 41, 61
[6] See, for example, Dreyfus H L 1979 *What Computers Can't Do* (New York: Harper and Row)
Born R (ed) 1989 *Artificial Intelligence: The Case Against* (London: Routledge)
Boden M (ed) 1990 *The Philosophy of Artificial Intelligence* (Oxford: Oxford University Press)
[7] For a history of the beginnings of Monte Carlo analysis see Cooper N G (ed) 1989 *From Cardinals to Chaos* (Cambridge: Cambridge University Press)
[8] Mihram G A 1972 *Simulation* (New York: Academic) p 209
[9] Heim M 1993 *The Metaphysics of Virtual Reality* (Oxford: Oxford University Press) p 89
[10] Husserl E 1970 *The Crisis of European Sciences and Transcendental Phenomenology* (Evanston, IL: Northwestern University Press) p 23
[11] Laszlo P 1993 *La Vulgarisation Scientifique; Que Sais-Je?* (Paris: Presses University de France) p 54
[12] Hilton J 1993 *New Directions in Theatre* (New York: Macmillan) ch 8
[13] Ringle M (ed) 1979 *Philosophical Perspectives in Artificial Intelligence* (Brighton: Harvester) p 14
[14] A digital computer was first used to study the dynamics of a large number of massive bodies—the *n*-body problem—by Sebastian von Hoerner in 1960 (von Hoerner S 1960 *Z. Astrophys.* **50** 184–214). Earlier, Erik Holmberg in 1941 used a novel analogue device to calculate the dynamical effects of a close encounter between two galaxies (Holmberg E 1941 *Astrophys. J.* **94** 385–95).
[15] Born R (ed) 1989 *Artificial Intelligence: The Case Against* (London: Routledge) p 71
[16] Waldrop M M 1993 *Complexity* (New York: Touchstone) p 146
[17] Ringle M (ed) 1979 *Philosophical Perspectives in Artificial Intelligence* (Brighton: Harvester) p 31
[18] Pavis P 1993 Production and reception in the theatre *New Directions in Theatre* ed J Hilton (New York: Macmillan) ch 3
[19] Leibniz G W *Philosophical Writings* (London: Everyman's Library) pp xvii, 152
[20] Ringle M (ed) 1979 *Philosophical Perspectives in Artificial Intelligence* (Brighton: Harvester) p 66
[21] Ringle M (ed) 1979 *Philosophical Perspectives in Artificial Intelligence* (Brighton: Harvester) p 233
[22] Ringle M (ed) 1979 *Philosophical Perspectives in Artificial Intelligence* (Brighton: Harvester) pp 234–5
[23] Barr A and Feigenbaum E A (ed) 1981 *The Handbook of Artificial Intelligence* vol 1 (Reading, MA: Addison-Wesley) p 180

[24] Dennett D C 1984 Cognitive wheels *Minds, Machines and Evolution* ed C Hookway (Cambridge: Cambridge University Press)
See also Thagard P 1993 *Computational Philosophy of Science* (Cambridge, MA: MIT Press)
[25] Shank R C 1990 What is AI anyway? *The Foundations of Artificial Intelligence* ed D Partridge and Y Wilks (Cambridge: Cambridge University Press) p 12
[26] Dewey J 1985 *How We Think and Selected Essays 1910–1911 (The Middle Works of John Dewey 6)* ed J A Boydston (Carbondale, IL: Southern Illinois University Press) p 61
[27] Dewey J 1985 *How We Think and Selected Essays 1910–1911 (The Middle Works of John Dewey 6)* ed J A Boydston (Carbondale, IL: Southern Illinois University Press) p 205
[28] Smith Churchland P 1989 *Neurophilosophy* (Cambridge, MA: MIT Press) p 407
[29] Beard N 1990 Mind over micro *Personal Computer World* **13** 182–6
[30] Crick F 1989 The recent excitement about neural networks *Nature* **337** 129–32
[31] Goldman A I 1986 *Epistemology and Cognition* (Cambridge, MA: Harvard University Press)
[32] Chalmers A 1990 *Science and its Fabrication* (Milton Keynes: Open University Press) p 13
[33] Koestler A 1989 *The Ghost in the Machine* (London: Arkana) p 195
[34] Barthes R 1986 *The Rustle of Language* (Oxford: Blackwell) p 69
[35] Plato *Theaetetus* 155d
[36] Voss H-G and Keller H 1983 *Curiosity and Exploration* (New York: Academic) p 150
[37] Berlyne D E 1960 *Conflict, Arousal and Curiosity* (New York: McGraw-Hill)
[38] Williams B 1987 *Descartes* (Baltimore, MD: Penguin) p 282
[39] Toates F M 1980 *Animal Behaviour—A Systems Approach* (New York: Wiley)
[40] Toates F M 1980 *Animal Behaviour—A Systems Approach* (New York: Wiley) p 222
[41] Toates F M 1980 *Animal Behaviour—A Systems Approach* (New York: Wiley) p 223
[42] Toates F M 1980 *Animal Behaviour—A Systems Approach* (New York: Wiley) p 225
[43] Peirce C S 1958 *Collected Papers* vol VII (Cambridge, MA: Harvard University Press) p 42
See also Habermas J 1972 *Knowledge and Human Interests* (London: Heinemann) p 133
[44] Vernon M D 1971 *Human Motivation* (Cambridge: Cambridge University Press) p 20
[45] Vernon M D 1971 *Human Motivation* (Cambridge: Cambridge University Press) p 23
[46] Vernon M D 1971 *Human Motivation* (Cambridge: Cambridge University Press) p 25

[47] Sturrock J (ed) 1979 *Structuralism and Since* (Oxford: Oxford University Press)
[48] Barthes R 1986 *The Rustle of Language* (Oxford: Blackwell) p 5
[49] White R Autonomy as foundational *Questioning Foundations (Continental Philosophy V)* ed H J Silverman (London: Routledge) p 81
[50] Barthes R 1986 *The Rustle of Language* (Oxford: Blackwell) p 100
[51] Lévi-Strauss C 1977 *Structural Anthropology* vol I and II (Baltimore, MD: Penguin)
[52] Piaget J 1971 *Structuralism* (London: Routledge and Kegan Paul) p 5
[53] Boden M A 1979 *Piaget* (Brighton: Harvester) p 19
[54] Rosen H 1985 *Piagetian Dimensions of Clinical Relevance* (New York: Columbia University Press) p 5
[55] Piaget J 1971 *Biology and Knowledge* (Edinburgh: Edinburgh University Press) p 362
[56] Seltman M and Seltman P 1985 *Piaget's Logic* (London: Allen and Unwin)
[57] Boden M A 1981 *Minds and Mechanisms* (Ithaca, NY: Cornell University Press) p 236
[58] Kitchener R F 1986 *Piaget's Theory of Knowledge* (New Haven, CT: Yale University Press) p 211

Chapter 7

Theories and questions

7.1 Quantum mechanics

Before a scientific theory can be visualized in terms of a network of questions, there remain a number of problems to be addressed and analysed.

The primary one is that of translation. More specifically, how exactly does one translate concepts, formulae, hypotheses and experimental facts into questions and answers. This problem merely reveals the shortcomings of how we think about and represent the world. Words such as 'fact' or 'hypothesis' inescapably trap us in the Cartesian dualist world of subject and object, matter and mind. No fact, as such, is without its presuppositions. Even a simple measurement of length involving a ruler carries implications about the rigidity of the ruler and the object being measured. There are no facts without hypotheses and these categories do not divide science in a way which is necessarily useful in exposing the validity or otherwise of our theories. The key in making the ideas of a theory clearer is to first identify that which is to be translated. In order to expand and detail these ideas I consider some examples of physical theories with quantum mechanics the starting point.

Of all the major theories of theoretical physics, quantum mechanics is probably the most successful and one of the most contentious. Even in its early days, the problems associated with the theory were overtly philosophical as well as scientific; although this distinction seems to be more important today than in the past. Quantum mechanics is not just one theory but a group of theories all arising from and closely connected with the original classical theory. The various sub-theories often sport three letter acronyms such as QCD (quantum chromodynamics) and

QFT (quantum field theory) and are all concerned with phenomena on an atomic scale.

Quantum mechanics arose out of a problem in accounting for the equilibrium distribution of electromagnetic radiation in a hollow cavity (black-body law). The existing and well-confirmed electrodynamic theory predicted an energy proportional to the square of the frequency of radiation (the Rayleigh–Jeans law) [1]. At low frequencies this corresponded very well to the experimental results but at high frequencies the Rayleigh-Jeans law predicted divergently higher energies than were observed. According to the Rayleigh-Jeans law the energy of radiation observed should increase for higher frequencies without limit, whereas the experimental results showed a maximum in the energy followed by a decrease at very high frequencies. Thus, at each temperature a black-body will have a peak value in the distribution of frequency of radiated energy. At a simple level an iron bar heated up will grow red and then as it gets hotter, its colour will change towards a 'white heat' until it melts. There were two problems, firstly, how to account for the experimental results and secondly, why the existing theory should predict energies which though accurate at low frequencies should exhibit an infinite divergence for high frequencies—a prediction which was non-physical regardless of the experimental results. This state of affairs led Planck in 1901 to propose the restriction that all allowed energy values were in discrete quantum units, proportional to the frequency, and this resulted in a new (Planck) law and the birth of a theory.

A second problem (the photoelectric effect) was considered by Einstein in 1905. Experiments showed that electrons emitted from a metal surface which was irradiated with ultraviolet light behaved as if the incident light energy was in discrete units proportional to the frequency, the constant of proportionality having the same value as that appearing in the Planck theory.

These were clearly two problems in one. Firstly, whereas electrons could be considered as particles, the corpuscular theory of light was long dead and buried. This followed all the pioneering theoretical and experimental work in the nineteenth century which established the wave theory of light. The photoelectric effect thus seemed to fly in the face of the idea of light as a wave phenomenon. Secondly, the fact that the same physical constant (Planck's constant) appeared in two diverse experimental situations seemed more than a coincidence and presaged some deeper connection.

A third problem was the model of the atom proposed by Rutherford in 1911, which failed both to explain the patterns of spectral lines previously discovered and how a negatively charged electron maintained a stable orbit around a positively charged nucleus to which it was strongly attracted. In this model the atom was pictured as a kind of solar system in miniature with 'planetary' electrons orbiting a nucleus. Calculations of the expected stability of the 'orbit' of an electron showed that the atom would disintegrate almost instantaneously. Since ordinary matter seemed perfectly stable, there was obviously a problem with the theory.

These three problems led Bohr in 1913 to propose the quantization of atomic electrons to discrete energy levels involving the ubiquitous Planck constant—this solved all three problems. The word 'solved' is perhaps going too far. But the new picture of the Bohr atom invoked the quantum concept which was applicable to a wide range of problems. The subsequent discoveries and refinements of the theory brought out the quantum concepts of wave–particle duality, indivisible quanta, quantum uncertainty and wavefunctions.

This brief perusal of early quantum theory is not meant as an historical précis, but highlights the problem-generating scenario involved in the genesis of the theory. The way I have set out the résumé makes the task of translation quite straightforward.

Suppose I number the questions as follows:

(1) How do you explain the black-body spectrum?
(2) How do you explain the photoelectric effect?
(3) How do you explain the presence of discrete spectral lines?
(4) How do you explain the stability of the atom in the Rutherford model?
(5) How does the quantization of electrons relate to energy quantization?
(6) What is the nature of the Planck quantum of action?

and the answers thus:

(a) The energy is quantized.
(b) The electron is quantized.
(c) The Planck constant.

The resulting graph representing early quantum theory is shown in figure 7.1.

This directed graph has six vertices of which four are leaves. All four leaf vertices have arrows pointing away from the vertex

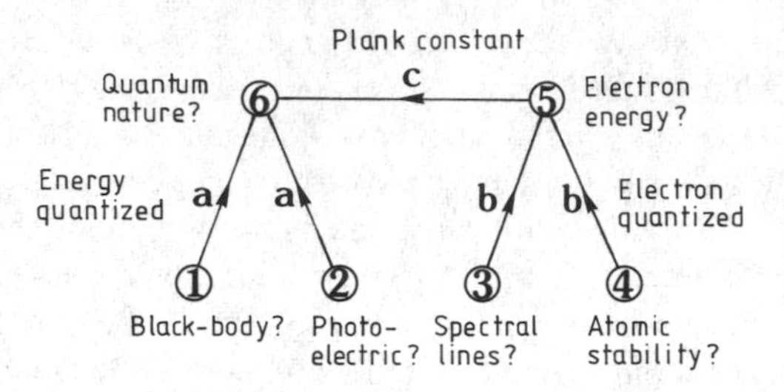

Figure 7.1 Digraph for early quantum theory.

(out-degree = 1, in-degree = 0), and the graph is acyclic; it is in fact a directed tree. Since it topologically conforms to all the rules of the zetetic model it is therefore an acceptable theory and so its success, until replaced, is deserved.

Note that question (4) highlights the problematic nature of the Rutherford model of the atom, even though the details of the model itself do not appear explicitly in the graph. The model, even though it is a theoretical construct, is taken as a given empirical problem within the zetetic network of the improved Bohr model. In a pictorial and practical sense the Bohr model has swallowed the Rutherford model. The other empirical questions (1) to (3) also contain elements which have been condensed or 'black-boxed', i.e. black-body, photoelectric effect, atomic spectral lines. Each of these phenomena can be represented by their own network of questions and answers arising from experimentation, theoretical speculation and analysis. But within the present context of the Bohr model, these networks have been suppressed and represented as empirical inputs into the model. The only questions exhibiting sub-detail are numbers (5) and (6) which refer to concepts and their relations arising within the model, and are thus classed as theoretical. We can see at once why the Bohr model was so successful even without constructing the question-and-answer framework. My main point is that by casting the experimental and theoretical aspects into an answer–question scenario, the validation of the model can be clearly seen in terms of the combinatorial structure.

7.2 Relativity

In 1905, when Einstein's first paper on special relativity appeared, there was a profound problem in theoretical physics concerning whether absolute velocity was measurable. The problem was both at the experimental and theoretical levels. At the theoretical level the two great theories of Newtonian classical mechanics and Maxwell's electrodynamics were incompatible. Classical mechanics implied that velocities were relative and that it was not possible to measure an absolute velocity. However, Maxwell's theory treated electromagnetic phenomena as the propagation of waves through some medium—the aether—and hence according to that theory the velocity of the Earth could be measured absolutely with respect to the aether frame.

Experimentally the situation was confronted by Michelson and Morley in 1887, who attempted to measure the velocity of the Earth through the aether by comparing the speed of light in two perpendicular directions. They found that no absolute velocity could be detected and the velocity of light was independent of the velocity of the reference frame.

Something had to go and Einstein's contribution was to overthrow the Newtonian concepts of space and time. This profound revision of concepts which had been central to physics for generations enabled Maxwell's theory, which united light and electromagnetism, to be consistent with the null result of the Michelson–Morley experiment. Newtonian classical mechanics was replaced with a new theory—special relativity—and the aether concept was demolished [2, 3]. The thrust of Einstein's argument in 1905 was to show that the impossibility of measuring an absolute velocity and the independence of the velocity of light from the velocity of its source were not logically incompatible with each other, provided one introduced a new transformation law between inertial frames. This transformation law—the Lorentz transformation—enabled Maxwell's laws to hold in their basic form regardless of the uniform motion of objects, whereas the former Galilean transformation changed the form of Maxwell's equations when objects were moving at constant velocity. This latter situation was not consistent with experimental observations and so either Maxwell's equations or the Galilean transformation had to be dropped. The Galilean transformation represented simple addition of velocities which was the common-sense view. However, this view was incompatible with Maxwell's equations. Einstein dropped the 'common-sense' view, or rather showed that a modified view of common sense was possible.

The background to Einstein's theory involved the special nature of

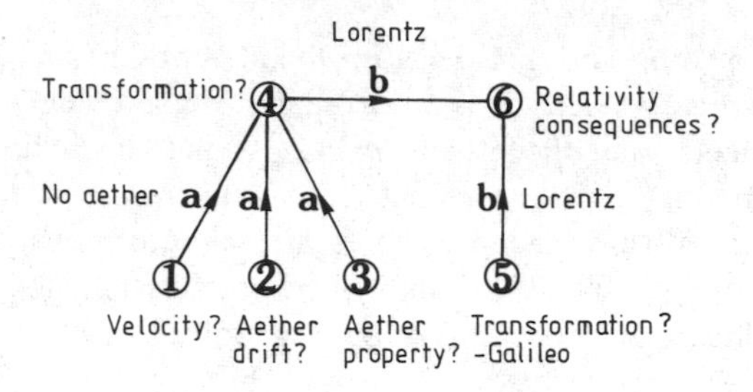

Figure 7.2 Digraph for relativity.

light and its propagation. Einstein considered that if he were seated on a train in a station looking at the station clock, then the time shown on the clock was visible as a result of the rays of light from the hands of the clock travelling at the speed of light to his eyes. If his train then moved out of the station then Einstein conjectured that as the train picked up speed it would catch up with the rays of light and the clock would appear to slow down, assuming it was still visible. Beyond the speed of light, if the train could go that fast, the clock would appear to be running backwards and so would all the people on the station platform. In this thought experiment, the Einstein on the train would even be able to see himself boarding the train 'in reverse'! In his youth this paradox preyed on Einstein's mind and led directly to his later theoretical ideas.

In terms of the zetetic scenario consider the following questions:

(1) Is velocity with respect to the aether measurable?
(2) Why is no aether drift measured in the Michelson–Morley experiment?
(3) Can the aether concept be dispensed with because its putative properties are so strange?
(4) If there is no aether and the velocity of light is a universal concept, what transformation applies to inertial frames?
(5) Is the Galilean transformation correct?
(6) What are the consequences of special relativity?

and answers:

(a) There is no aether.
(b) The correct transformation is that of Lorentz not Galileo.

The graphical representation of Einstein's theory in 1905 is shown in figure 7.2.

As in the example of early quantum mechanics, the digraph in figure 7.1 shows a valid structure, though of a different topological character. The point about special relativity as an example is that it involved the opening-up of a black-box. Maxwell's and Newton's theories were both valid and acceptable theories at the time and had been black-boxed, but because they were incompatible one had to be modified or abandoned—the step taken by Einstein.

In this and the quantum-mechanical example I have only sketched over the historical detail. My purpose is not to analyse the theories and the circumstances of their discovery, but to abstract a way of looking at scientific theories in terms of questions and answers and to demonstrate this process of abstraction and translation.

At the same time, part of the present model involves a different way of perceiving scientific theories. Obviously to a practicing scientist who may be in the middle of performing some experiment it is hard to immediately translate what is being done in questioning terms. However, at the theoretical, or even at the 'take a step back' level, to see or analyse what is happening in terms of questions posed is not difficult, although it may require a change in viewpoint. It is this change in viewpoint which is the central theme of this work.

The zetetic model will not, however, legislate between two valid models. Nor will it guarantee that answers in a network are either correct or even relevant. The validity is only in the combinatorial structure and this is the result of the way the theory is constructed by scientists. The questions and answers which make up the framework represent problem areas and the way they are related to each other. The actual language or the precise form or wording of a question or answer is unimportant. What counts are the problems or problem areas as perceived and the answers, or alternatively 'relata', which are applied.

7.3 Zetetic analysis

I have described the two theories of early quantum mechanics and relativity in very simplistic terms. There is, of course, a great deal more to these theories than the rather bald historical recreation that ninety odd years of hindsight gives us. But the advantage of choosing examples of this type is that the underlying 'zetetic' structure is more easily shown. What is omitted from the present analysis is all the arguments, discussions and other ideas (or questions) which were around when these theories were in their infancy, together with all the experimental work and interpretations which made up the scientific background both then and now. I utilize the hindsight we have in this generation to filter out all these complexities and concentrate on the hard-core, established or 'black-boxed' parts of these theories. The theories are not dead but are part of the present scientific canon and so are thus recreated in their present form and structurally analysed in this way.

There is also a dynamical aspect of theory change. From the inception of a theory relevant questions or problems are either included or omitted, as the case may be. In general terms, the changing structural pattern of a theory is represented by transformations of structure mediated by structural rules. Patterns which remain static for a long enough time become black-boxed. We can never say that the final pattern emerges as black-boxes can always be opened. However, the two examples employed in this chapter represent very good cases of theories, or at least parts of theories, which have become static in the above sense.

One of the purposes in using these two particular examples is to show as starkly as possible the relationship between questions and theories and how a theory acquires the description 'scientific'.

Most of science research goes on at the sub-theory level. This is meant in the sense that it is rare for a new theory to be put together in one go and remain intact (structurally speaking) for a considerable time afterwards. Individual researchers may not be consciously or even unconsciously building complete theories but are just attempting to fit a small number of pieces together within some framework—they try to create modules rather than models.

There is, of course, no way to predict how theories will change and when. No matter how stable a black-box is, there are always unanswered questions and it is these, in particular, which leave the situation open. In addition, it is always an option for a black-box structure to be dismantled and reassembled in some different form. Beyond the structure of a theory, the open-endedness means that there will always be problems in fitting a theory into the larger scheme of things. By this I mean that ultimately there is no such thing as an autonomous theory. This is not to say that only a holistic approach would be successful, in fact a truly holistic approach is equally unattainable in practice. Even today after nine decades of quantum theory and relativity, the precise status of, for example, photons is still very much under discussion.

Even though the two examples of theories considered in this chapter are both established, black-boxed theories, the zetetic method of analysis provides a number of useful insights to them as well as theories in general.

Firstly, even though we are recreating the theories by virtue of historical hindsight, by identifying the crucial questions and their relationships to each other, we can reconstruct the important developmental stages of a theory.

Secondly, the scientific aspects of a theory can be identified in a clearer way and alternative theories can be assessed in structural terms. This gives an added dimension to theories allowing us to discuss them in a language that describes their formal structure.

Thirdly, open-ended questions can be identified and areas which could be linked to other theories can be exposed more readily. This facilitates the setting of a theory within a wider context.

Fourthly, from the philosophical or epistemological point of view, there is a representation of a theory at a structural or formal level rather than in terms of the contents or entities created by a theory.

Fifthly, it is quite useful to identify what is not a theory or at least what is close to being but is not quite a theory. These structures represent modules or sub-theories which may possibly be grouped into theories, but which fall short of the necessary characteristics of a scientific theory on their own. Many of these structures represent unsolved problems which because they are not part of an established theory lie dormant only to be resurrected at a later time when they can be usefully employed. The change of importance of these types of problems reflects not only their intrinsic importance but whether they can enhance a theory by their membership within the structure.

The two theories exemplified in this chapter have some points in common which are brought out by zetetic analysis of their structures. Both cases involve six questions of which four are 'empirical'. If one ignores the labelling of the graphical structures of the two theories shown in figures 7.1 and 7.2 and just concentrates on the topological features, then there are only three distinct digraphs which are allowed under the zetetic rules. The three possible digraphs consist of the two depicted in figures 7.1 and 7.2 plus a third represented by the graph in figure 7.2 with the direction of the arrow between vertices 4 and 6 reversed. Thus from a structural point of view there is a well-defined and strictly limited number of topologically distinct 'scientific' digraphs that can represent a theory.

Notes and references

[1] Bohm D 1951 *Quantum Mechanics* (Englewood Cliffs, NJ: Prentice Hall)
[2] Angel R B 1980 *Relativity: The Theory and its Philosophy* (Oxford: Pergamon)
[3] Rosser W G V 1964 *An Introduction to the Theory of Relativity* (London: Butterworths)

Chapter 8

Theories in the making

8.1 Dark matter

The examples I have considered so far—quantum mechanics and special relativity—are well established theories although they do contain problematic areas and will no doubt be replaced by better theories in the future. The rather narrow areas of these theories considered in the previous chapter are firmly black-boxed and, as I have already mentioned, tend to be thought of almost as deductive areas of a theory—unless the black-box is opened. Failing this, the problematic areas move higher up the hierarchy away from the black-box areas, and new theories tend to take over these black-box areas *en bloc*.

The zetetic model so far emerges more as a picture theory of science than as an analytical tool. Also, the fact that in the zetetic model these well-established theories are seen as valid is no more than a necessary conclusion given the way in which the structures of the theories are represented. My purpose in these examples is to highlight the process of translation whereby theories may be represented as a question and answer framework.

To take a well established theory and re-represent it in such a way to show that its structure is that of a well-established theory is not of itself great progress. However, in the next examples I want to analyse theories which are either not well established or even not well formed. In this latter sense I refer to the flotsam and jetsam of new ideas and anomalous observations which make up the semi-conscious sub-culture of science. These include fledgling theories and bits of theories. They are like the various sections of an unfinished jig-saw puzzle: some may consist of a few pieces fitted together, others may be known to be pieces of sky, for example, and yet others may be from a different puzzle. The first example is referred to as the dark matter or missing mass problem [1].

There are really two problems that come under this heading. The first is a discrepancy in the estimated masses of galaxies measured using two different methods. The second problem is a discrepancy between the global density of the universe thrown up by certain preferred cosmological models and the value of this parameter as measured by astronomers.

The first problem came to light originally in the 1930s when Fritz Zwicky studied the dynamics of the Coma cluster of galaxies. Our own solar system together with some several thousand million other stars is part of a galactic structure or 'island universe'. Many other similar such galaxies are observed by astronomers and it is found that galaxies themselves often congregate into bound groups or associations referred to as clusters. The hierarchical ordering in the universe is thus: stars, galaxies, clusters of galaxies, super-clusters and the universe. (Note the additional level of clustering between clusters of galaxies and the universe called super-clustering.)

The Coma cluster of galaxies studied by Zwicky is one of the largest clusters observed with a membership of about eight hundred galaxies covering an observed diameter of about four degrees in the sky. Zwicky found that the relative speeds of individual galaxies within the cluster were so high that given the total mass of galaxies present, the cluster would have dispersed in quite a short time. Since the cluster appeared to be a stable entity, Zwicky concluded that the luminous material in the cluster—the observed galaxies and gas—only constituted about one-tenth of the total mass. The missing mass dominated the dynamics of the cluster but was in a non-luminous form. Over the last fifty years many studies have been carried out which have corroborated Zwicky's original findings.

In general terms, if any objects are gravitationally bound to each other in some sort of stable or quasi-stable system and if we have information on the distance and speeds at which the objects are moving, then we can elicit information on the masses of the objects. In the case of clusters of galaxies things are not perfectly straightforward—they rarely are in astronomy. Apart from uncertainties over distances, the velocities which are measured by spectroscopic analysis only refer to projected velocities along the line of sight to the cluster and not to the actual velocities. In spite of all this uncertainty, we can home in on mass values for clusters and what emerges from these studies is that the masses of clusters are much higher than expected from simply counting up the masses of individual galaxies which appear to make up the cluster.

That the amount of unseen matter in clusters of galaxies is ten times the amount of visible matter is strange enough, but the second problem implies that the amount of matter in the universe is *one hundred* times more than the visible matter. This additional tenfold increase in the amount of matter in the universe over and above that thought to reside in clusters of galaxies, is implied by acceptance of a particular cosmological model within the standard Big-Bang model of the universe.

To follow the steps leading to the preferred model, it is necessary to go back over the chequered history of cosmological models since Einstein first proposed his theory of general relativity in 1916. The famous field equations gave the mathematical structure for the dynamics of the universe; however, there were a number of free parameters and hence different available cosmological models within relativity. A particular set of models were developed by A A Friedmann in the early 1920s, and later by Robertson and Walker, and applied to the general class of isotropic, homogeneous universes with zero cosmological constant. (The latter cosmological term was introduced by Einstein into the field equations as a means of obtaining a static universe. After Edwin P Hubble's discovery in the late 1920s that the universe was expanding, Einstein abandoned the cosmological term [2].) Since on the largest scales the universe is observed to be isotropic and homogeneous, the Friedmann models are the ones most studied.

In the Friedmann models, the Hubble constant, H_0, which is the rate of expansion of the universe, together with the gravitational constant, G, the constant of proportionality in the inverse square law of gravitation, generate a critical density parameter ρ_c given by:

$$\rho_c = \frac{3H_0^2}{8\pi G}.$$

The present density of the universe is thus some fraction, Ω_0, of this critical density. In physical terms the critical density may be thought of as a balance between the explosive forces which constitute the expansion of the universe compared with the attractive force of gravitation. A universe with a density at the critical value ($\Omega_0 = 1$) corresponds to one where these attractive and repulsive forces exactly balance. The geometry of the universe depends on whether the density is lower, higher or at this critical value. If the present density of the universe actually corresponds precisely to the critical value then the universe will expand forever—but only just. A lower density signifies an open model where expansion wins over gravity and a higher than critical density results in a closed model for the universe which will reach a maximum size and then eventually re-collapse on itself.

The Friedmann–Robertson–Walker models provide for any value of Ω_0. Up to a few years ago, apart from aesthetic reasons for believing that Ω_0 should be unity, as opposed to some other value, the cosmologists aim was to measure Ω_0 rather than invoke theoretical values.

It might appear a little strange that aesthetics or simplicity should be a factor in science, but there is a profound tradition going back to ancient Greece which sees harmony and simplicity as part of nature. Ockhams Razor dictates that if Ω_0 is known to lie between 0.1 and 1.5, then it is probably unity. But Ockham's Razor, despite its superficial attraction, has been greatly overused. Named after William of Ockham (c1285–1349) and quoted (though not by Ockham himself) as 'entities are not to be multiplied beyond necessity', it is an important principle of economy.

What is really behind the attraction for this value of Ω_0 is that it corresponds to a universe whose total dynamical energy exactly balances its gravitational energy, and it is this natural harmony which seems to be more than just a coincidence, but an indication that there is more theory yet to be found which will point to this value of Ω_0. What really changed the situation was a theoretically predicted value for Ω_0 of precisely unity, which arose out of a theory called inflationary cosmology.

The theory itself evolved as a result of a number of problems associated with the standard Big-Bang model of the universe. The first two of these problems involved what was not seen by astronomers rather than what was seen. In a Big-Bang cosmology the universe was much smaller and hotter in the past. The universe cooled down as it expanded and today its temperature is close to three degrees above absolute zero. This can be measured by observing the background microwave radiation which is a relic of the former hot phase of the universe. After allowing for local and galactic distortions this radiation is found to be consistently equal in all directions in the sky. This observed isotropy of the microwave background radiation implies that in the early stages of the universe, regions which were totally unconnected causally all exhibited precisely the same temperature at the same time. The problem here is that even though the microwave background radiation may have started out in the early universe highly constant in all directions, subsequent non-thermal events such as galaxy condensation and formation of larger scale structure would distort the isotropy so that the effective temperature of the radiation should be slightly different in different directions. This variation is not observed.

The second problem is that the value of Ω_0 is unstable, in the sense that an extremely small departure from a value of unity in the past would result in a value very different from unity at the present time. The magnitude of this effect, referred to as the flatness problem, is to constrain the value of Ω at the earliest stages of the universe to be closer to unity than one part in 10^{59}. Expressed in a slightly different way: if Ω was not equal to or incredibly close to unity in the early universe then by now its present value would be much further away from unity than observations allow.

The third problem is to explain the mass concentrations which we observe in the universe: galaxies, clusters, super-clusters and voids. Given an expanding universe and the known laws of physics, the scales of matter condensations should arise naturally as a result of the growth of density perturbations and correspond with observed structures. What actually happens is that there is no particular preferred scale of clustering, and the inferred density perturbations in the early universe, which would grow to form galaxies and clusters, are seen as arbitrary initial conditions which have to be plugged into the model.

These problems, among others, were neatly addressed by work in the 1980s on inflationary universe scenarios. In these models the description of the universe back to within about 10^{-30} seconds of the initial singularity agrees precisely with the standard Big-Bang model. However, for times earlier than this, the universe went through a brief period of exponential expansion. During this inflationary period the universe expanded by about a factor of $\sim 10^{50}$ [3].

There are a number of different versions of the inflationary scenario but they all follow broadly the same pattern. By introducing a period of exponential inflation in the evolution of the universe, all inhomogeneities decay exponentially and so the extreme isotropy of the microwave background radiation is explained. The model also predicts $\Omega_0 = 1$.

As far as galaxy formation is concerned, since inflation stretched microscopic scales into astronomical ones, galaxy formation could

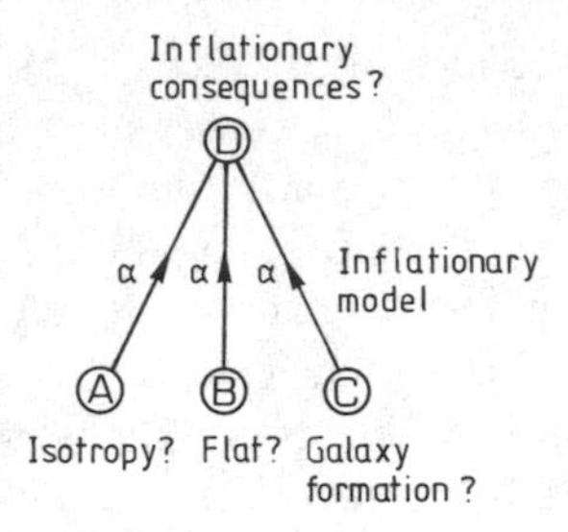

Figure 8.1 Digraph for the inflationary universe model.

result from quantum fluctuations; the observed matter condensations would then depend on the amount of dark matter in the universe. In this way the inflationary scenario obviates the need for arbitrary initial conditions.

The rather superficial résumé which I have given does not address a number of problems in the models, but since I have deliberately chosen an example of a theoretical structure in the making, it is no surprise that there are very few areas that have become black-boxed.

To consider the inflationary scenario first there are three main problems, namely:

(A) Why is the microwave background radiation so isotropic?
(B) Why is the universe so flat?
(C) Why do the density perturbations in the early universe, which give rise to galaxies etc have such arbitrary values?

These can all be accommodated by the answer:

(α) The early universe undergoes a period of 'inflationary' expansion.

The graph for the inflationary scenario has the simple form shown in figure 8.1.

The question (D) generated by the scenario is:

(D) What are the consequences of an inflationary model?

The graph in figure 8.1 does not, of course, represent the whole picture. Firstly, I have left out other problems which the scenario solves, e.g. the non-observance of magnetic monopoles and, secondly, the details of the inflationary model give rise to other problems, some of which have led to the model being radically revised, though the basic paradigm remains.

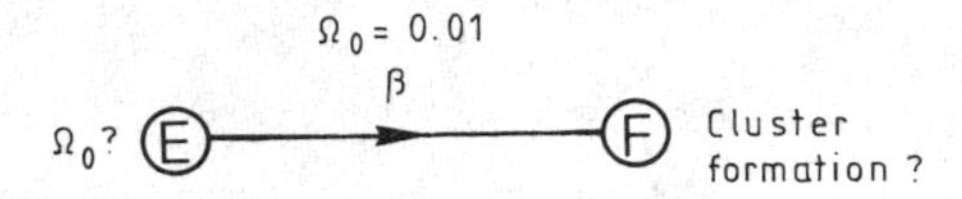

Figure 8.2 Digraph for the dark matter model, $\Omega_0 = 0.01$.

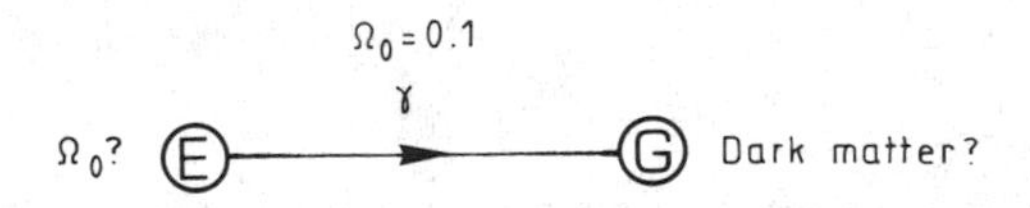

Figure 8.3 Digraph for dark matter model, $\Omega_0 = 0.1$.

Figure 8.1 represents how the model is viewed by cosmologists. The power of the idea of inflation is that it can cope with three problems which are currently considered important.

Moving from the inflationary scenario to the dark matter problem we can add the following questions:

(E) What is the value of Ω_0?

(F) Why is the mass density of visible matter insufficient to stabilize clusters of galaxies?

(G) What does the dark matter consist of?

and the following answers:

(β) $\Omega_0 = 0.01$

(γ) $\Omega_0 = 0.1$

(δ) $\Omega_0 = 1$.

We can now look at the graph structures generated by the answers β, γ and δ.

Figure 8.2 represents the initial stalemate caused by the apparent contradiction between the observed and dynamically inferred masses of clusters.

Figure 8.3 is equally unhelpful. It solves the cluster stability problem but at the expense of another problem, namely that of dark matter.

Figure 8.4 shows the full structure. The three cosmological problems (A), (B) and (C) are solved by the inflationary scenario which implies a dark matter problem, but within an acceptable theoretically valid framework. This is the reason why the dark matter hypothesis is considered seriously by astronomers.

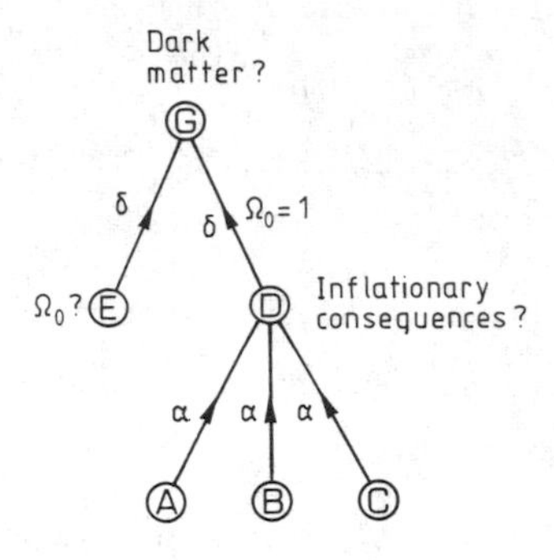

Figure 8.4 Digraph for dark matter model, $\Omega_0 = 1.0$.

Of course, figure 8.4 does not show the whole picture, eventually the theory will be supplanted by another by the process of continual scrutiny, criticism and observation. However, at this stage the graph represents how a scientific theory or scenario can cope successfully with a number of problems, and in order to amend or destroy the theory an equally valid structure must be built. The robustness of the network as measured by the difference between the number of empirical and theoretical questions (in this case two) represents the resistance to change of a theory, and is a measure of how far the theory has been black-boxed.

At this stage the main question is:

(G) What does the dark matter consist of?

In the theory so far we have moved up the hierarchy to the problematic area of trying to answer this question.

There are a number of answers to this question at the moment, none of them yet regarded as definitive [4]. The two kinds of matter postulated to make up 'dark matter' are labelled 'hot' or 'cold'. This refers to the velocities of the types of particles; either hot (at or close to the speed of light, e.g. neutrinos) or cold (at low velocities, e.g. ordinary matter, axions and photinos). The reason for this broad division is that the two types of matter produce structures in the universe in different, (and hence possibly observable) ways. Hot matter forms structures on the largest scales which in turn condense to form clusters and galaxies in a 'top down' hierarchy. Cold matter, on the other hand, first forms small-scale clumps which act as seeds for galaxy formation in a 'bottom up' scenario.

At first the cold matter idea had the upper hand because it was

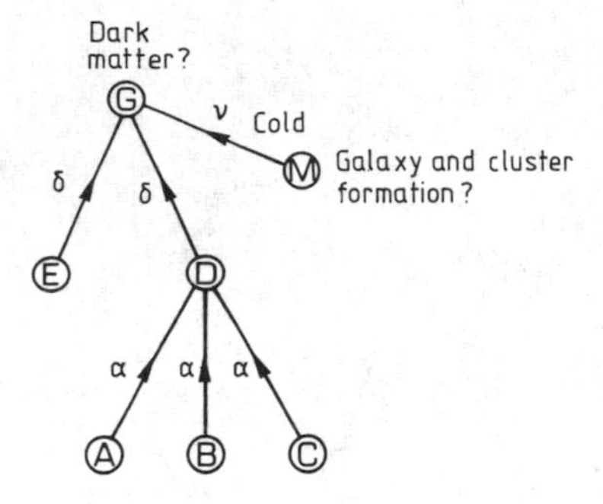

Figure 8.5 Digraph for the cold dark matter model before the results of the far-infrared survey.

successful in accounting for the formation of galaxies and clusters in a way which was consistent with observations. However, observations of a survey of galaxies carried out in the far infrared showed very large-scale structure in the universe which was much lumpier than predicted by the cold matter theory [5].

Let us go back to the stage just before the results of this far infrared survey were known. We can pose the following question:

(M) How do galaxies and clusters form?

and answer:

(ν) Dark matter is 'cold'.

The resultant graph is shown in figure 8.5, where the existing theory covers question (M) thereby strengthening the whole scenario.

But after the results of the far infrared survey had been published, the situation changed because of the very large-scale structure which the cold dark matter hypothesis had difficulty in accounting for.

If we add the following question:

(N) How does very large-scale structure form?

then the cold dark matter theory is pictured as in figure 8.6.

In terms of the difference in number between empirical and theoretical questions, which in the present model means internal and leaf vertices respectively, a 'gain' of three (figure 8.5) is reduced to two (figure 8.6). Furthermore if question (N) 'How does very large-scale structure form?' is considered to be of greater importance, and hence greater weight, than other questions, the gain in figure 8.5 will be less than two.

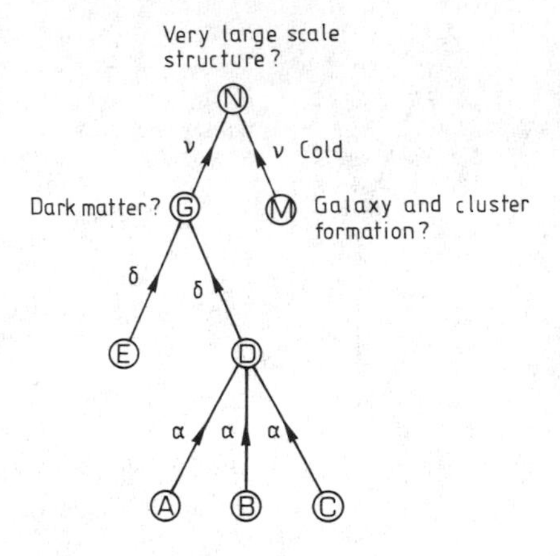

Figure 8.6 Digraph for the cold dark matter model after the results of the far-infrared survey.

In view of this it is no surprise that other theories have appeared (and no doubt others will follow) which threaten to replace or significantly modify the cold dark matter theory. I consider two theories which have been proposed.

Firstly, there is a suggestion by George Efstathiou and two colleagues that a cosmological model involving a non-zero cosmological constant might be a useful modification to the cold dark matter theory [6]. I have earlier in this chapter referred to the cosmological term and its somewhat chequered history. Because of this there is a general unwillingness to introduce it into current theories. This is based, no doubt, on the fact that cosmological theory and fundamental physics have been black-boxed so as to exclude this term, and there is a natural reluctance to open black-boxes without good reason. Efstathiou and his colleagues have resurrected this term (designated Λ_0) for two reasons. Firstly, they can keep all the good features of the original model, namely the inflationary scenario in the very early universe and the isotropy of the microwave background. Secondly, the need for dark matter is obviated because the dynamical effects needed are provided in effect by the cosmological term. Recall the answer (γ) in relation to the density parameter:

(γ) $\Omega_0 = 0.1$.

In the light of the non-zero cosmological constant model, I introduce a modified answer:

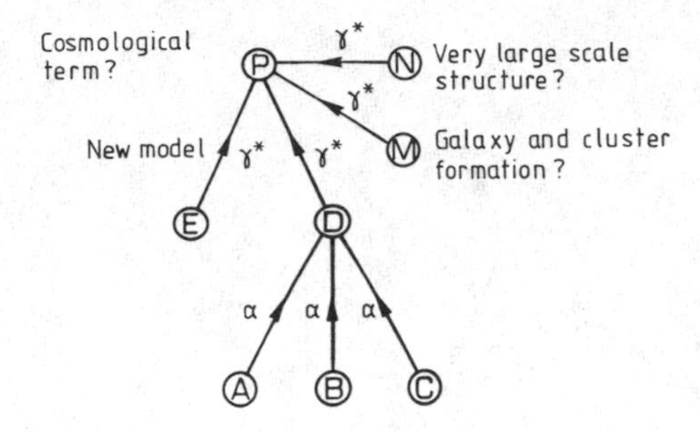

Figure 8.7 Digraph for the non-zero Λ_0 model.

$(\gamma^*)\ \Omega_0 = 0.2 \quad \Lambda_0$ non-zero.

This establishes the hypothesis of the new model, and I have doubled Ω_0 to 0.2 in order to bring it into line with the model of Efstathiou *et al* [6]. In cosmological terms the factor two is not important.

If we further introduce a new question:

(P) What are the consequences of a non-zero cosmological term?

then we can draw the graph representing the new model shown in figure 8.7.

The gain of this model is four, representing a significant improvement on the previous model. Of course, the question (P) 'What are the consequences of a non-zero cosmological term?' shifts the argument to this level of the hierarchy and the model will be severely tested on this point.

The second alternative model which I should like to consider has been put forward by Dennis Sciama [7]. His approach is to postulate that dark matter consists of neutrinos. The neutrino was first proposed by Wolfgang Pauli in 1931 to explain discrepancies in the process of beta decay where an atomic nucleus emits or captures an electron. Part of the reaction involves a neutrino which interacts via the weak nuclear force. There are in fact three types of neutrinos (and anti-neutrinos), each associated with one of the three known leptons: the electron, the muon and the tau.

There are also problems associated with neutrinos in an astrophysical context. Because neutrinos via the beta process are involved in nuclear reactions in the core of the Sun, their flux can be measured on the Earth. Over the past twenty years or so, various experiments have been set up to do this. One thing these experiments consistently have in common—

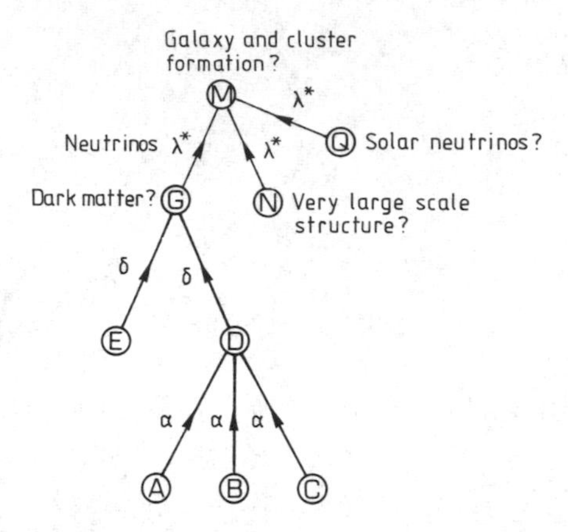

Figure 8.8 Digraph for the neutrino model.

and which is the source of the problem—is that the observed flux of neutrinos is about a factor three or four times lower than that expected from predictions of the standard solar model.

One explanation of this discrepancy involves neutrinos having non-zero rest masses, a consequence of which is that they may supply the missing mass.

Figure 8.8 shows the graph representing Sciama's model. A new question (Q) has been covered by the scheme:

(Q) How do you explain the lower than expected flux of solar neutrinos?

and a new answer (λ^*) has been introduced:

(λ^*) Dark matter is in the form of massive neutrinos.

In Sciama's model the question of galaxy formation is still problematical, but the gain of the theory is three, and for those who consider the solution of the solar neutrino problem highly important, the model has much to commend it. Naturally if the mass of the neutrino is established experimentally, then the gain of Sciama's model will increase dramatically.

So far I have considered three different models each being valid from the zetetic model point of view and each covering slightly different problems. As I have explained before, the zetetic model will not legislate as to which model is the right one—the concept of 'right one' does not apply. They are all in a sense right and which will fall quickest by the wayside will depend on future observations and theorizing.

8.2 Anthropic principle

In the forward to Barrow and Tipler's *tour de force*, *The Anthropic Cosmological Principle* [8], John Wheeler wryly states that 'it is often more difficult to ask the right questions than to find the right answers, and nowhere more so than in dealing with the anthropic principle' [9]. In referring to both the content and status of the principle, he was highlighting both scientific problems as well as scientists' prejudices.

In its weak form, Barrow and Tipler argue that 'the observed values of all physical quantities are not equally probable but they take on values restricted by the requirement that there exist sites where carbon-based life can evolve and by the requirement that the universe be old enough for it to have already done so' [10]. As Arthur Schopenhauer puts it in his book *The World as Will and Idea* [11]: 'If in general there is to be a world at all, if its planets are to exist at least as long as the light of a distant fixed star requires to reach them ... then certainly it must not be so clumsily constructed that its very framework threatens to fall to pieces'.

There are a number of different assertions here apart from the fairly trivial observation that the existence of observers such as ourselves imposes some selection effects on what we see around us [12]. Implicit in the definition is the idea that in our universe, which is a unique entity, the various physical constants could have had different values. This reflection arises more out of ignorance than theory. Because we have no theory which says that, for example, the mass of the proton must be 1840 times the mass of the electron, we therefore consider that this number is a value among a range of possible values for that constant, and the range of possible values is limited by the selection effect caused by virtue of our own existence. Of course, in another universe vastly different life forms may have evolved and the same musings might be made, although involving a completely different range of possible values of physical quantities.

The anthropic principle, however, goes further than a sterile debate on selection effects, it asserts that different values of physical quantities are not only possible, but that our universe is only one of an infinite number of universes each with its own set of laws and with different values for physical quantities.

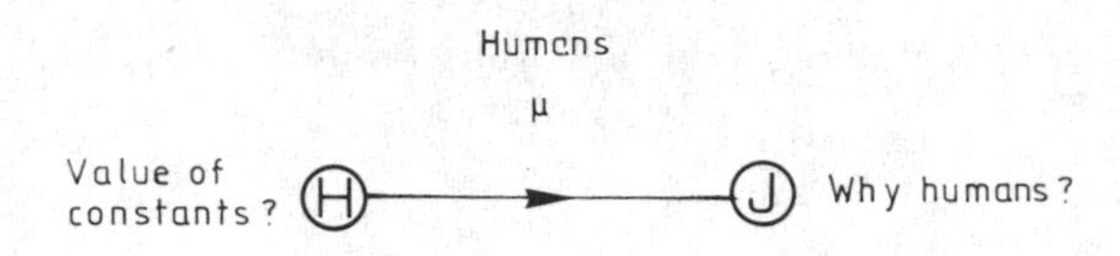

Figure 8.9 Digraph for the strong anthropic principle.

This idea goes well beyond the view of induction as hypothesizing about possible states of affairs. There is nothing wrong with thought experiments so long as they can be performed in principle. The anthropic principle when confronted with a thought experiment whose in-principle performance is in doubt, boldly changes a possibility into a certainty.

The problem with invoking other universes is that the explanatory value of theoretical postulations normally derives from some causal link between the entity postulated and the natural world. However, in the case of other universes, they come under the class of non-natural *possibilia* which are semantic postulations in no way explaining anything that happens in the natural world [13]. In the zetetic model, however, we take the practical view that if a postulate answers a question and forms part of a valid graph structure, then it counts as a theory.

Consider the following questions:

(H) Why do the dimensionless constants in the universe have the values they do?
(I) How do universes with different dimensionless constants form?
(J) Why do humans exist?

and answers:

(ϵ) Within limits dimensionless constants can have any value in different universes.
(μ) Because human beings exist.

The basic question (H) can be answered in two ways. First there is the strong version of the anthropic principle [14]. This states that the universe must be the way it is because of the existence of human beings. The graph for this is shown in figure 8.9.

A homeomorphically similar graph is obtained for the 'weak' anthropic principle as shown in figure 8.10.

Both represent invalid or incomplete structures. However, in the previous section on the dark matter hypothesis when we considered the

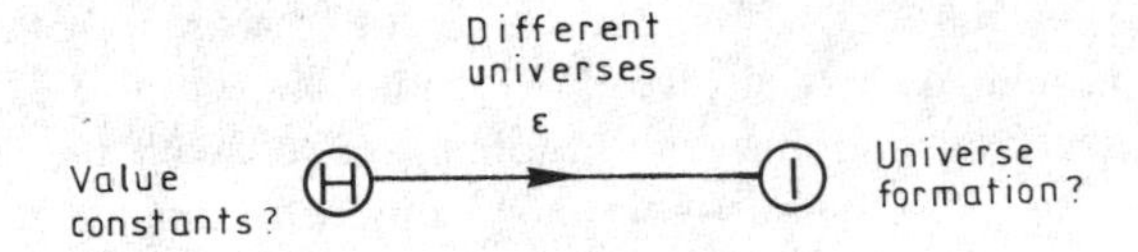

Figure 8.10 Digraph for the weak anthropic principle.

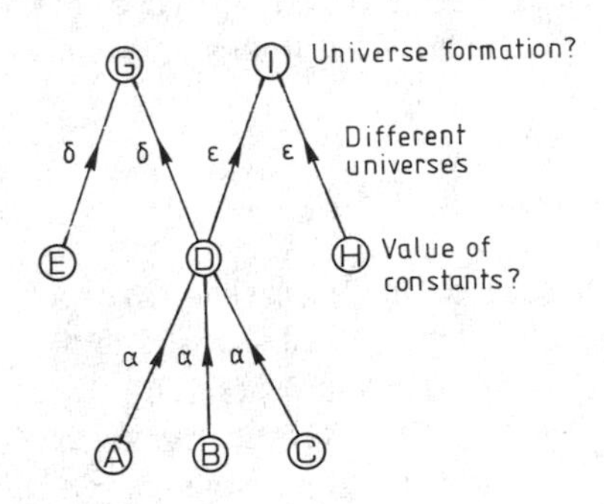

Figure 8.11 Digraph for the combination of the weak anthropic principle and the dark matter hypothesis.

inflationary scenario, one of the consequences of this model is that the quantum fluctuation which could give rise to our universe could also give rise to other universes with alternate values for physical quantities.

Recall question (D) 'What are the consequences of an inflationary model?' and the graph figure 8.4.

We are now able to add the anthropic principle to the graph and obtain figure 8.11.

This shows the changed character of the weak anthropic principle from an invalid or incomplete piece of theory to a part of a valid structure which is encompassed within the inflationary scenario.

Furthermore, because the addition of the anthropic principle adds one internal vertex and one leaf vertex, it does not change the gain of the theory. Thus the anthropic principle can also be added to each of the three models which I considered in the previous section of this chapter. In a more general way we can state that:

A theoretical answer/question can always be introduced into a theory provided it is attached to another empirical question.

How the theory will develop in the future one cannot say at the moment. All the above questions, together with others in quantum theory and cosmology, form the basis of research both at the experimental and theoretical level [15]. Either the graph will enlarge as further questions and answers are amalgamated with the structure, or the structure will break up—perhaps being reformulated in another way. The zetetic model is thus a representation of the dynamics of theory change both as evolution and revolution. It is thus more than just a picture theory of science but an analytical tool which can show the changing rational structure of a theory.

8.3 Theory of everything (TOE)

In his inaugural lecture to the Lucasian Chair of Mathematics at Cambridge, Stephen Hawking posed the question as to whether the end was in sight for theoretical physics [16]. By 'the end' he was referring to a long-standing research programme initiated by Einstein to unify the known laws of physics into one theory fully encompassing both quantum mechanics and general relativity. To Hawking, this end is only a first step in the goal of a complete *understanding* of the events around us and of our own existence [17, 18]. Is this the fulfilment of Socrates' tongue-in-cheek view [19] of:

> *those who inquire into celestial things, imagine that, when they have discovered by what laws everything is effected, they will be able to produce, whenever they please, wind, rain, changes of the seasons, and whatever else of that sort they may desire(?)*

But what does it mean to reach even this first step? The implication behind the putative unification is not just that an equation will be found from which the known laws of nature and forces will emerge, but that this equation and the theory behind it will remain unchanged. The problem here is that so long as science is viewed as some Cartesian object under study with truth as a goal, then contradictions will arise. It may very well be that an equation will arise in the future which will be labelled the unification equation, but its status will be no different from Galileo's equation relating the distance an object falls under gravity (d) to the time of fall (t)

$$d = \tfrac{1}{2}gt^2$$

where g, the gravitational acceleration, is a constant, or Einstein's famous equation for mass (m) and energy (E) ($E = mc^2$, where c, the velocity of light, is a constant). These equations are important and well known, and no doubt will remain within science but they do not in themselves constitute theories. Theories are concerned with the process of science and the never-ending supply of questions about nature which scientists weave together into theories. These theories, or parts of them, may be black-boxed for a time, but so long as there is the possibility of opening the black boxes then understanding can never be complete; you can always ask one more question: Why?

Of course, why stop at physics, why not a theory of everything, resulting in the end of philosophy? This particular conundrum has already been considered by Heidegger under the title 'The end of philosophy and the task of thinking' [20]. He considered two questions:

> (1) *What does it mean that philosophy in the present age has entered its final stage?*
> (2) *What task is reserved for thinking at the end of philosophy?*

To the first question, Heidegger interprets the final stage as the dissolution of philosophy. However, he does not take the negative implication of this which suggests an end or decline to completion or perfection. On the contrary, it is the development of the sciences and their autonomous separation from philosophy which moves the latter towards dissolution. This is an on-going process which in the past has resulted in physics, chemistry and mathematics taking over areas which used to be the sole province of philosophy. In addressing the second question, Heidegger puts forward the ideas of openness and 'unconcealment' underpinning the task of thought to 'what is', rather than 'what is demonstrable'.

Clearly, philosophy still has a long way to go with plenty of life after the end. Looking down the line at physics, which in the above scheme is an autonomous branch of philosophy, the 'what is' and the 'what is demonstrable' would seem to be just as much in evidence both now and after a TOE is put forward. As far as the zetetic model is concerned, so long as a TOE satisfies the combinatorial criteria for its constituent questions, then it is an acceptable theory. And if after a number of theories are put forward there comes a time when the whole area can be black-boxed, then what will interest philosophers and scientists will be the questions generated—that will be the task for TOE 2!

Notes and references

[1] Excellent discussions of this topic are to be found in:
Rees M J 1987 Galaxy formation and dark matter *300 Years of Gravitation* ed S W Hawking and W Israel (Cambridge: Cambridge University Press)
Krauss L 1989 *The Fifth Essence* (London: Hutchinson, Radius)
Gribbin J and Rees M J 1990 *The Stuff of the Universe* (London: Heinemann)

[2] Pais A 1982 *Subtle is the Lord* (Oxford: Oxford University Press)

[3] Blau S K and Guth A H 1987 Inflationary cosmology *300 Years of Gravitation* ed S W Hawking and W Israel (Cambridge: Cambridge University Press)
Linde A 1987 Inflation and quantum cosmology *300 Years of Gravitation* ed S W Hawking and W Israel (Cambridge: Cambridge University Press)

[4] Rowan-Robinson M 1991 Dark doubts for cosmology *New Scientist* March 30–4

[5] Saunders *et al* 1991 The density field of the local universe *Nature* **349** 32–8

[6] Efstathiou G, Sutherland W J and Maddox S J 1990 The cosmological constant and cold dark matter *Nature* **348** 705–7

[7] Sciama D W 1990 Consistent neutrino masses from cosmology and solar physics *Nature* **348** 617–8

[8] Barrow J D and Tipler F J 1986 *The Anthropic Cosmological Principle* (Oxford: Clarendon)

[9] Barrow J D and Tipler F J 1986 *The Anthropic Cosmological Principle* (Oxford: Clarendon) p viii

[10] Barrow J D and Tipler F J 1986 *The Anthropic Cosmological Principle* (Oxford: Clarendon) p 16

[11] Schopenhauer A 1886 *The World as Will and Idea* (Engl. Transl. R B Haldane and J Kemp) (London: Trübner) vol III p 393

[12] Gribbin J and Rees M 1990 *The Stuff of the Universe* (London: Heinemann) p 287

[13] Armstrong D M 1989 *A Combinatorial Theory of Possibility* (Cambridge: Cambridge University Press)

[14] Barrow J D and Tipler F J 1986 *The Anthropic Cosmological Principle* (Oxford: Clarendon) p 21

[15] Linde A 1987 Inflation and quantum cosmology *300 Years of Gravitation* ed S W Hawking and W Israel (Cambridge: Cambridge University Press) ch 13

[16] Hawking S W 1980 Is the end in sight for theoretical physics? *Inaugural Lecture to the Lucasian Chair of Mathematics (Cambridge, 1980).* Reprinted in Boslough J 1989 *Stephen Hawking's Universe* (London: Fontana)

[17] Hawking S W 1988 *A Brief History of Time* (London: Bantam) p 169

[18] For an interesting and entertaining historical account of scientists who considered the proposal that science might discover everything, see Gardner M 1985 *The Whys of a Philosophical Scrivener* (Oxford: Oxford University Press) ch 20

[19] Xenophon *Memorabilia of Socrates 15* as published in *Plato and Xenophon. Socratic Discourses* (London: Dent) p 4

[20] Heidegger M 1972 The end of philosophy and the task of thinking *On Time and Being* (Engl. Transl. J Stambaugh) (New York: Harper Torchbooks)

Chapter 9

Darwinian evolution

9.1 Introduction

So far I have concentrated on examples of scientific theories taken from the physical sciences. This is a natural choice for me as my background is in astrophysics. Moreover, the physical sciences are perceived as the model for scientific thought at its most basic. However, one of the tenets of the present work is to view science in a much more general way as one aspect of an intellectual process, and as such it is important to apply the zetetic model in another area. I now turn to the biological sciences and in particular to evolutionary theory—the most important theoretical structure of biology. There is also an additional relevance to the present model which fits in with the evolutionary theory of knowledge. This is much more than an argument by analogy (see Chapter 6). In this chapter I consider Darwin's theory, neo-Darwinism and also an hypothesis put forward by Dawkins which goes under the rubric of the 'selfish gene', the title of Dawkins' book [1].

9.2 Darwin's theory

There is an enormous complexity to all living things both in the vast number of different species on Earth and in the life of each as it passes from birth to death. Yet, there appears to be some sort of a consistency of form from generation to generation for each species—individuals may come and go but the species continues. In ancient times some Greek writers considered the idea that species of living organisms could change over long periods of time. However, the immutability of species

Figure 9.1 Charles Darwin. (Reproduced by permission of Mary Evans Picture Library.)

was part of the biblical account of creation in the book of Genesis and was not questioned seriously until the eighteenth century.

According to Genesis, all species were created by a catastrophic rather than an evolutionary process and species did not evolve or develop into each other. This follows the literal account of the creation in the first two chapters of Genesis. However, in a later chapter (Genesis, Chapter 36:24) there is mention of a mule—a cross between a stallion and a she-ass—so the biblical account does include hybridization.

Historically, we have to distinguish between two usages of the word evolution. One refers to the development of an organism, and the other more modern usage refers to specific instances of transmutation as well as to the process of change in general [2]. The cycle from birth to maturity and reproduction is for many organisms one of major evolution of form—from the humble cabbage white butterfly and the frog to humans. However, regardless of the overwhelming changes of form that an organism may undergo during this process, the cycle is faithfully repeated through the generations, so that the idea of species or fixed types could be retained.

The classical view started to be seriously questioned in the eighteenth century by Buffon, Lamarck and others. Maupertuis, for example, in the mid-eighteenth century suggested that all present species were derived from a small number, or perhaps a single pair, of original ancestors [3].

The questions raised were part of a tradition whose roots were in antiquity and through the ages there were many 'partial anticipations' of Darwinian evolution. The idea of species as natural kinds which generally breed true, gave way to the eighteenth century idea of a species as a fixed type separated by the barrier of sterility. Even examples of hybrids such as the mule, mentioned in Genesis, were themselves sterile.

The first theory of the transmutation of organic forms was by Lamarck in 1800. In his theory simple organisms arose from spontaneous generation, but these successively produced or evolved into all the other species. The changes necessary were brought about by the environment and each organism's innate drive to adapt.

In Lamarck's system, environmental change led to an organism adopting new habits which in turn led to changes in the animal's structure, which were passed on to successive generations. Lamarck's system was the inheritance of acquired characteristics. The most famous example was the long neck of the giraffe. This was acquired by successive generations of giraffes stretching their necks more and more in order to reach leaves which were only abundant at higher levels in the trees. If a giraffe spent its whole life trying to stretch its neck as much as possible, then its progeny would actually be born with a slightly longer neck. The only evidence in favour of Lamarck's theory was the structural similarity between living things. Many different species of animals have necks, all structured along the same lines but differing in size. Lamarckian evolution would of course require much longer times than the biblical age of the Earth, namely six thousand years, and Lamarck among others believed that the Earth's age was very great.

It was the impetus from the geologists which set the stage for evolutionary theories. The Earth was thought to have been involved in prolonged and violent upheavals with major changes of sea level over very great periods of time measured in hundreds of millions rather than thousands of years. By studying fossils laid down in geological strata a picture was developing of different stages of Earth's history with widely different climates and flora and fauna, most of which are now extinct.

The genesis of Darwin's theory was his visit to the Galápagos Islands in September 1835. This was part of a five-year voyage in the *Beagle*, and Darwin collected an enormous number of different species, particularly finch and turtle, which inhabited the islands, each apparently filling a particular ecological niche of their own. After his return to England he developed his theory and explained such diversity by postulating a common ancestry from some original finch or turtle which had come to the islands from the mainland.

He had in 1837 started his first 'Notebook on the Transmutation of Species' and the final ingredient to his ideas was provided by a reading of Malthus' work on population in 1838 [4]. In any species of organism the adults tend to replace themselves by a greater number of progeny. There is thus a vast potential for increase in all populations. An organism which doubles its number in a year would undergo a million-fold increase in twenty years. This natural and exponential increase in population occurs within an environment of resources which have some natural limit of capacity. The result is that population pressure will always give rise to competition amongst organisms. Thus, if any variant form amongst a population possesses some competitive advantage over the rest of the population, it will tend to survive and reproduce at the expense of the rest of the population.

Darwin's essential ingredient which was the basis of this theory was the mechanism by which species originate, namely 'natural selection'. This idea of adaptive change, independently put forward by Wallace, was published by Darwin in 1859. Wallace had reached his ideas whilst laid up with a fever in the Moluccan islands. Spurred on by reading Malthus and his own socialist background he wrote to Darwin in 1858 of his theory of natural selection, just as Darwin was finishing his long-gestated manuscript [5]. Wallace's ideas were slightly different from Darwin's on a number of points, one of which was that Wallace saw the environment as eliminating unfit organisms rather than competition among individuals being the selection mechanism. However, the letter galvanized Darwin into action and he went public with his ideas.

In the revolutionary climate of the 1840s, Malthus' ideas were used to justify the reduction of government aid to the poor and combat the political demands (including universal suffrage) of many reformers [6]. In this background, evolutionary theory became synonymous with riot and rebellion and Darwin was reluctant to make his ideas public during this period—discretion being the better part of valour. His work was not published until the end of the 1850s.

Darwin put forward a wealth of evidence for his theory, developing a complete biological system of adaptive change which was the result primarily, though not exclusively, of natural selection operating over long periods on the small variations present in animal and plant populations.

The first part of the evidence for evolution is the geographical distribution of animals and plants. In particular in Australia which had been isolated from other land masses for around 60 million year, marsupial or pouched mammals are the only naturally occurring mammals. The fossil evidence indicates that 60 million years ago placental mammals did not yet exist.

Since that time, a large number of species of Australian marsupials have developed which bear great similarities to placental animals which have evolved elsewhere in the world. In spite of there being no close biological relationship there are pouched mammals which look like dogs, cats and even lions. For Darwin especially, the isolation of the Galápagos from the South American mainland and the myriads of separate species, often each to its own island, showed species evolution in action.

The second type of evidence which Darwin put forward and which covered a crucial assumption in his theory, was the variation present amongst populations of animals and plants. Clearly, there are variations in the sense that individuals in a population exhibit differences between each other in size, colour and other characteristics; however, it was variations which are passed on from one generation to the next in which Darwin was interested. Under domestication by man, many animals and plants have been selectively bred by propagating derived characteristics which magnify initially small natural variations. Examples are the diverse number of different breeds of dogs, which were ultimately derived from a single wild type and selective breeding of cattle with higher milk yields. Even though in a number of species sex ratios vary, it is interesting to note in this latter example that selective breeding of cows which only give birth to female calves would be advantageous to dairy farmers, however, all efforts to do this have failed. The reason, in modern terminology, is that there is no genetic variation for this particular characteristic, namely giving birth only to females. It is also not difficult to appreciate that even if only a single cow were present in a natural population, which could only give birth to females and which passed on this characteristic to her progeny, then within a small number of generations the whole population would become extinct.

The third type of evidence for evolution comes under the heading of morphology or anatomy. This more than any other evidence makes the separate creation of different species a totally unbelievable hypothesis. If you compare specific organs of animals of different species but which are member of the same class, for example, mammals, then there are homologous similarities. So that the arm of an ape, the leg of a dog, the flipper of a whale and the wing of a bat, in spite of all looking quite different externally, have an underlying skeletal plan which betrays a common ancestry [7]. There are something like a quarter of a million species of flowering plants but all (except for a few parasites) share the same basic and recognizable structure of roots, stem-bearing branches, leaves (with chlorophyll) and flowers. In the insect world the number of species is about three quarters of a million; however, all have a similar body pattern of head, thorax and abdomen with three pairs of legs and two pairs of wings. In fact the whole biological classification scheme right back to Aristotle picks out familial relationships between species, based on arguments from homology. This applies also to extinct fossil species.

Intercomparisons between species show even stronger similarities if instead of comparing the adult forms one looks at embryological development. Firstly, as one looks back to earlier stages of the development of a bat's wing or a whale's flipper, external differences appear much less and, secondly, an organ may start to develop at the embryo stage only to disappear or end up as a vestige or functionless organ in the adult. An example of the former is the gill slit in mammals which betrays a marine ancestry and an example of the latter is the appendix in man which serves no purpose, although in other species it can be an important organ for digestion. Similarities of form between species is also mirrored by similarity in behaviour patterns. Examples are nest building among birds and social behaviour in ants, bees and wasps.

In his famous book *The Origin of Species* [8] Darwin took great pains to discuss the problems and difficulties as well as the evidence in favour of his theory. A problem, even today, is the incompleteness of the fossil record. The famous 'missing link' *Archaeopteryx*, intermediate between lizards and bird, was first discovered in 1861, and in recent years bones and whole skeletons belonging to ancestors of *homo sapiens* have been discovered in Africa. As Darwin wrote [9]:

> *... there must have been a thousand intermediate forms between the otter and its land ancestor ... opponents will say, show me them. I will answer yes, if you will show me every step between the bulldog and greyhound.*

One problem for Darwin was the origin of highly complex structures, particularly the eye, which reportedly left him in a 'cold sweat' [10]. But evolution of complex forms as well as complex instinctual behaviour can be seen as a long series of gradations of complexity. Thus adaptive change is seen as the result of natural selection operating over long periods. This process depends on how much time there is for evolution of this kind to work, and thus depends on the geological timescale.

In Darwin's time there was a problem with the age of the Earth as estimated by geologists based on the cooling time of the Earth. (Darwin suggested 300 million years back to the dinosaurs [11].) Physicists, led by Sir William Thompson (Lord Kelvin) considered that the Earth could not have supported life for more than a few million years. The latter timescale was much too short for Darwinian evolution to have taken place. This particular problem was not resolved in Darwin's favour until after his death.

In order to analyse Darwin's theory we need to identify the basic questions which he was asking and the answers which provided the theory's structure. This is particularly easy in this case since Darwin's notebooks are full of questions and his whole approach was just that of asking questions.

We can start by summarizing five main questions as follows:

(1) How are species related to each other?

(2) Why do species have a common morphology?

(3) Why do species have embryological similarities?

(4) Why are there so many diverse species?

(5) Why is geographical isolation important?

These are all answered by:

(a) Species all have a common ancestry.

which is the first part of Darwin's theory. At this stage his theory is no different from other transmutation theories and leads to the crucial question:

(6) How do species evolve?

The core of Darwin's theory is the answer to this question and the identification of the actual mechanism of evolution. The answer provided by Darwin is:

(b) Adaptive selection of heritable variations over a long period of time.

However, this leaves the unanswered questions:

(7) What is the mechanism of heredity?

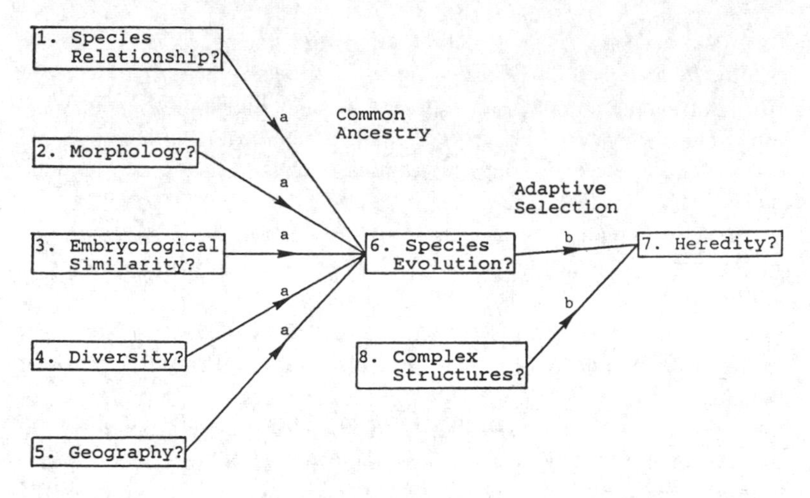

Figure 9.2 Digraph representing Darwin's theory of evolution.

(8) How do complex structures form?

Thus natural or adaptive selection is the missing piece of the jig-saw according to Darwin. Provided the timescale is long enough, randomly occurring small advantageous transmutations will accumulate so that through the generations organisms can efficiently fill biological niches which changing climate or landscape provide.

The graph in figure 9.2 shows clearly the two parts to the theory. The first part consists of questions (1) to (6) and the answer (a), and forms a very strong hypothesis consisting of five empirical questions covered by one theoretical question. This part of the theory merely asserts that the hypothesis of common ancestry to all species can answer a number of diverse empirical questions and is not original to Darwin. The second part of the theory is important as it covers the evolution of complex structures but does not improve the theoretical structure as it adds one empirical question and one theoretical question. Furthermore, the answer that Darwin provides, namely adaptive selection, depends on the geological timescale being very much longer than many of the physicists of his time would allow. Ultimately Darwin was proved right on the point but not until after his death.

It is very difficult to look back on a theory, particularly one such as Darwin's which is well accepted nowadays, and try to weigh the evidence as it was at the time in an objective way. Because of the timescale problem, which arose out of basic physics, going beyond 'common ancestry' to 'natural selection' did not necessarily improve the theoretical structure or give rise to a better theory. Of course, Darwin could claim (and did) that very long geological timescales were needed to lay down the layers of strata in which he and others observed fossils. This process also depended on basic physics of a more mundane but no less important kind.

However, greater corroboration of his theory, or in line with the present work. I should say a better zetetic graphical structure, developed afterwards.

Since Darwin's time his theory has developed into neo-Darwinism or the synthetic theory of evolution. This follows on from the work of the Augustinian abbot Gregor Mendel whose work though published in 1865, went unnoticed until 1900. Mendel addressed the problem of the mechanism of heredity, (question 8 on our list), and identified 'factors' (genes) in the reproductive cells of organisms which transmitted characteristics—thus genetics was born.

Heredity is seen in terms of the action of genes. We now have the complete hierarchical scale of sexual reproduction from the egg and the sperm right down to the molecular level. The basic structure of all organisms is the cell, with most bacteria and blue-green algae belonging to the oldest group of organisms which are without a nucleus and which go back to the dawn of life on Earth some three billion years ago. All other organisms, which have cell nuclei, do not seem to have been present on Earth before about two billion years ago.

Early microscopic studies of cells revealed filamentary units called chromosomes which were in the cell nuclei and which split and reformed when cells divided. These chromosomes carry genes which are made of deoxyribonucleic acid (DNA). The life of the cell is regulated by the DNA which carries the genetic information which is passed on to each cell. When cells are formed DNA is replicated and each cell acquires its own chromosome structure which enables the process to continue. There are two mechanisms of cell propagation, mitosis and meiosis. In mitosis, after a period of growth, a cell divides into two daughter cells each with a copy of the original DNA chromosome structure. Meiosis occurs in sexual reproduction: special cells are produced with half of the chromosome complement and these subsequently pair with other similar cells to produce cells with the full chromosome complement. The importance of meiosis from an evolutionary point of view is that all possible combinations and recombinations of genes can be realized. In mitosis all progeny have an identical genetic make-up whereas in meiosis all cells have a random permutation of chromosomes of equal maternal and paternal origin. If, for example, in a human being we focus on the DNA of one cell, (not a sex cell), then looking back on its life we would see a seemingly endless string of mitotic divisions interspersed every seven or so times with a meiotic division. Thus changes in the structure of chromosomes on replication bring about new permutations of genes which give rise to mutations. Natural selection leads to the elimination of mutant genes with harmful effects and encourages the spread of genes of favourable mutations in a population.

The source of variation in organisms is the result of the random mutation of genetic material. Sometimes this will produce markedly aberrant individuals which could give rise to sudden large changes or evolution by macro-mutation [12]. Also there are occasions when Lamarckian inheritance can occur, when acquired characteristics are passed on to progeny, but these do not involve any change in nuclear DNA itself. Rather, the genetic activation or even the number of genes change, and these effects pass through a number of generations [13].

However, the overwhelmingly important evolutionary mechanism is that of natural selection working over long periods on small Mendelian variations produced by mutation at the DNA level. Species are seen as sets of organisms which share a gene pool.

Species today have gene pools which consist of the selected and modified fractions of gene pools of more or less distant common ancestral species. Different species arise when parts of a gene pool become genetically isolated. This may occur as a result of geographic isolation, but whatever the mechanisms, since breeding success is much lower between hybrids, natural selection would favour organisms which tend to mate with members of their own species. What neo-Darwinism provides is the basis for heritable variation, namely random mutations of DNA material.

There is of course a vast wealth of material and discoveries which I have not touched on, the preceding discussion being only a brief thumbnail sketch of neo-Darwinism and genetics. The main point, though, is that most of basic genetics has been black-boxed. We can represent this by the following questions:

(9) How do cells divide and reproduce?
(10) What are chromosomes?
(11) What gives rise to heritable variation?
(12) What are genes?

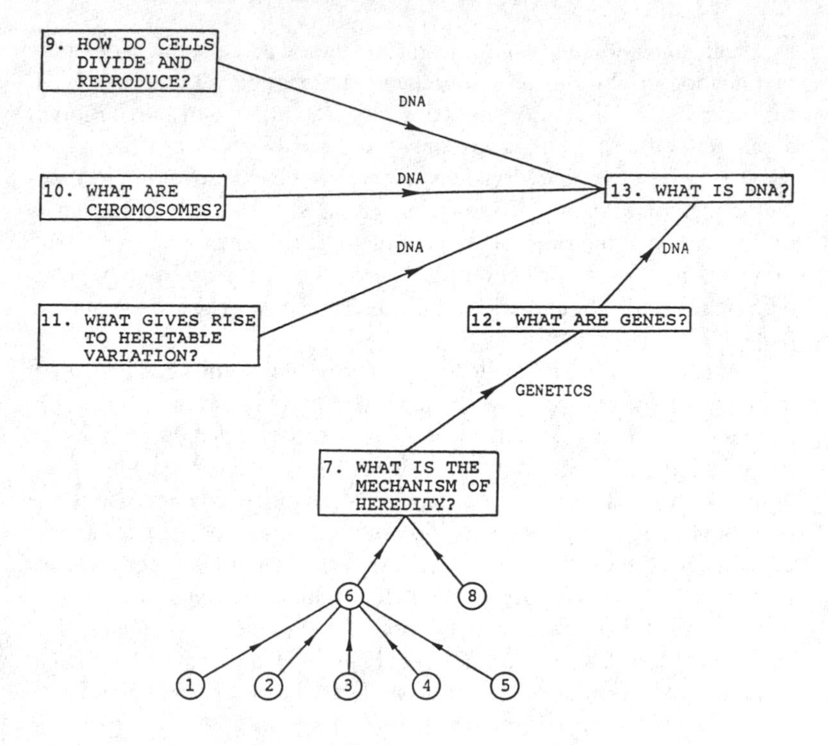

Figure 9.3 Digraph representing neo-Darwinism.

(13) What is DNA and how does it evolve?

We can then 'add' genetics to the original Darwinian model which is shown in figure 9.3.

We could replace the whole of figure 9.3 by a single empirical question:

(13) What is DNA?

which would represent the complete black-boxing of the whole theory and which represents for research workers within the field the question to which they are applying all their efforts. Of course, this is not to say that at some future time someone may come along and open up the black-box by questioning some basic tenet of the theory.

Some, in the past, have even gone as far as to say that the theory is tautological. Flew [14] has represented this situation by the graph shown in figure 9.4 where arrows represent logical implication.

Thus Malthus is represented by the geometrical rate of increase

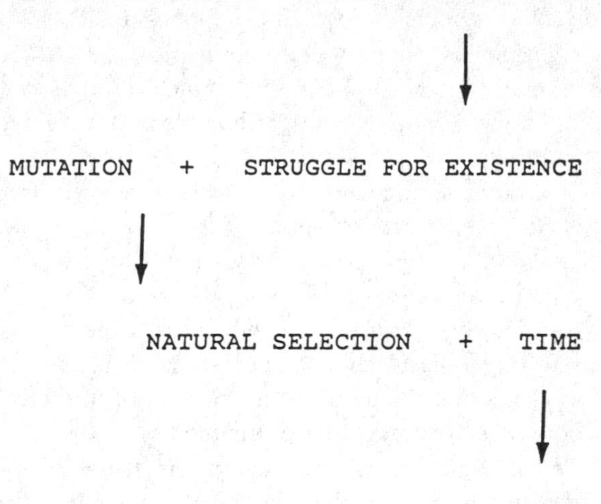

Figure 9.4 Diagram showing an 'implication digraph' for Darwinian evolution [14].

of organisms and limited resources being the core of a deductive argument which leads to a struggle for existence. Add to this mutation (or in the case of classic Darwinism, variation) and natural selection is a consequence. Given sufficient time this leads to biological improvement.

That this 'deductive' argument is false is shown best by Popper [15], who originally thought it was valid. Without natural selection an isolated population would exhibit 'genetic drift' but would not evolve apparently designed organs like the eye. Since not all organs serve a useful purpose, although most do, natural selection, though important, is not completely universal and the theory is not just a tautology.

9.3 The extended phenotype

In looking at organisms, genes, heredity and DNA we have an interlocking hierarchical system which Dawkins [16] has pointed out cannot be studied in separate parts. His basic idea is that the effects of a gene in an organism—the phenotype—is 'best seen as an effect upon the world at large, and only incidentally upon the individual organism—or any other vehicle—in which it happens to sit' [17].

A gene 'is not just one single bit of DNA ... it is *all replicas* of a particular bit of DNA, distributed throughout the world ...' and furthermore '... a gene might be able to assist *replicas* of itself which are sitting in other bodies' [18].

Dawkins summarizes his view in a central theorem: 'An animal's behaviour tends to maximize the survival of the genes "for" that behaviour, whether or not those genes happen to be in the body of the particular animal performing it' [19].

As Dawkins himself pointed out, his thesis 'may not constitute a testable hypothesis in itself, but it so far changes the way we see animals and plants that it may cause us to think of testable hypotheses that we would otherwise never have dreamed of' [20].

What Dawkins is emphasizing is the cooperation of cells and organisms for mutual benefit thereby extending the network to gene structures throughout populations and amongst species. In terms of the basic question posed

(14) Why do cells and organisms cooperate?

the answer is the same as to how species evolve and how complex structures evolve, namely by adaptive selection. At first sight the result of Dawkins' thesis, as far as the present model is concerned, is simply to append one empirical question onto the structure in figure 9.2. This takes into account the property that Darwinian evolution can account for evolution of complex structures both at the organism level as well as at the level of assemblies of organisms.

I say at first sight, because one could argue that question (14) is already part of the theory, embedded in question (8) and that very little empirical content has been added to the theory. From the point of view of the zetetic framework, Dawkins' ideas therefore are fairly neutral. They can thus represent a useful and novel approach to evolutionary theory without in themselves being important as far as the validity of the theory is concerned.

This is not to decry Dawkins' hypothesis, after all most philosophical ideas and theories are in the same boat, in that they do not necessarily constitute scientific theories. My point is that the demarcation between science and non-science is more clearly seen in terms of the relations between the questions generated.

Notes and references

[1] Dawkins R 1976 *The Selfish Gene* (Oxford: Oxford University Press)
[2] Desmond A and Moore J 1991 *Darwin* (London: Michael Joseph) p 40
[3] Flew A 1984 *Darwinian Evolution* (London: Paladin) p 13
[4] Flew A 1984 *Darwinian Evolution* (London: Paladin) p 7
[5] Desmond A and Moore J 1991 *Darwin* (London: Michael Joseph) p 468
[6] Desmond A and Moore J 1991 *Darwin* (London: Michael Joseph) p 315
[7] See 1991/92 *Pears Cyclopedia* (London: Pelham) F.30
[8] Darwin C 1859 *The Origin of Species* ed J W Burrow (Baltimore, MD: Penguin)
[9] Desmond A and Moore J 1991 *Darwin* (London: Michael Joseph) p 248
[10] Desmond A and Moore J 1991 *Darwin* (London: Michael Joseph) p 320
[11] Desmond A and Moore J 1991 *Darwin* (London: Michael Joseph) p 566
[12] Sheldrake R 1987 *A New Science of Life* (London: Paladin) p 142
[13] Smith J M 1989 *Evolutionary Genetics* (Oxford: Oxford University Press) p 11
[14] Flew A 1984 *Darwinian Evolution* (Paladin) p 37
[15] Popper K R 1977 Natural selection and its scientific status *A Pocket Popper* ed D Miller (Fontana) p 242
[16] Dawkins R 1982 *The Extended Phenotype* (Oxford: Oxford University Press)
[17] Dawkins R 1982 *The Extended Phenotype* (Oxford: Oxford University Press) p 117
[18] Dawkins R 1982 *The Extended Phenotype* (Oxford: Oxford University Press) p 153
[19] Dawkins R 1982 *The Extended Phenotype* (Oxford: Oxford University Press) p 233
[20] Dawkins R 1982 *The Extended Phenotype* (Oxford: Oxford University Press) p 2

Chapter 10

Literature and science

10.1 Introduction

To view science as a human response to the natural world by the creation of an intellectual weave also allows for other kinds of intellectual activity which, though not science, could and should be modelled in the same way. Certainly language, as discussed in Chapter 6, falls into this category. However, literature, or more particularly fictional literature does not appear, at first sight, to bear much structural or basic relationship to science. I believe that this harsh division between literature and science is partly illusory and in this chapter I shall enlarge on this point.

There are of course, many areas of interest between science and literature where the demarcation between science and non-science is blurred. This is, I believe, as it should be since there are activities which in themselves may be science or non-science depending on the approach. The aim of a pursuit may not be the construction of a scientific theory, but it may be, for example, to effect political change or the financial management of a large organization. To say that the resulting politics or economics is not scientific, using the term in a derogatory sense, is misguided. Politics and economics look at the world in different ways from science and the rules with which they are constituted are consequently different.

Let me state first of all my two aims: firstly, to show that there is an underlying unity between literature and science. This unity is not only at the superficial level of language—which is common to both— but that both reflect a basic human trait of the way we cope, adapt and look at the world. My second aim is to highlight the liberating aspects of this unity. I refer here to elitism or the 'ivory tower' syndrome which serve to enslave science as an enterprise that is far removed and cut off from the rest of the intellectual family. By seeing science as just another kind of intellectual pursuit, albeit involving manipulation of the world, there is a greater freedom of action, particularly in crossing into areas formerly considered off-limits to science. I concentrate in this chapter on literature and science, but the unity to which I am referring applies to all intellectual pursuits.

10.2 Fictional entities [1]

Let me start with a story. Yesterday, I met Mary Poppins. Perhaps I should explain a little more. I was talking to the concierge at our hotel in Disneyworld—my children insisted that we included this stop on our holiday in the United States—when an exotic figure appeared in my peripheral vision. My eyes did a double-take but my mouth just carried on talking, although I had no idea what I was saying. The concierge seemed to notice my reaction and she informed me:

It's Mary Poppins.

I reacted with some expression of disbelief but she continued:

Have you ever met Mary Poppins?

Still overcome by amazement, I managed to indicate that I had not, at which point she called loudly across the hotel foyer to the figure:

Oh, Mary Poppins, come over here and say hello to Nigel, he wants to meet you!

The figure turned round, fixing us with a permanent smile, and with a swish of her umbrella—it was well over 85 degrees Fahrenheit and dry outside the hotel—she briskly approached us.

'How do you do, Nigel' she said as I shook hands with her. After a suitably polite but almost inaudible reply from myself, we were quickly surrounded by a pack of children eagerly seeking autographs and taking pictures. After a few minutes she waved goodbye and disappeared from view. I turned back to the concierge who said to me in a conspiratorial voice:

You met Mary Poppins!

Well I suppose in a way I did. But on the other hand, surely the only person I met was an actress playing the role of Mary Poppins. Mary Poppins is not a real person but a fictional character. The reality of the situation is the person I met—whoever she was—and not the role which she was playing. As Jean-Paul Sartre (figure 10.1) explained particularly in his play 'Kean' [2], to refer to an actor by the part that is being played is an act of bad faith. Even as we go about our daily lives, we may be guilty of this bad faith by actively trying to play a part rather than being ourselves. The waiter who assiduously clears the tables and deftly sets out the cutlery (an example used by Sartre) is trapped in a role which limits his freedom of action to a point that he is guilty—according to Sartre—of bad faith; he is playing the role and not being the person. Consider further the plight of Kean [3]:

KEAN: *There was nobody on stage. No one. Or perhaps an actor playing the part of Kean playing the part of Othello. Listen—I am going to tell you something. I am not alive—I only pretend.*

And what about the plaintive opening lyrics of the song in *Blood Brothers* [4] sung by the tragic heroine:

Tell me it's not true, say it's just a story.

We know that it is a story, but nevertheless within the performance there is a certain veracity. After all, fiction depends on the truth for its effect.

A similar idea can be found in comedy. During a television interview with Roseanne Barr by Terry Wogan, when shown an excerpt from her comedy series and asked the question: Is that you? Roseanne replied [5]:

Yes, it's me when I'm acting; this is me as I am.

We have here different levels in a hierarchy. Roseanne Barr's role as a comedy actress in a series, as interviewee and perhaps her role or roles as a person. For, in a sense, we are all in part at least the sum of all the roles which we play. Obviously in a theatrical setting on the stage, film or television the assumption of a role by an actor is deliberate, involving an artistic aim with an audience in mind. To describe someone as playing a role in their everyday life is slightly different, we refer instead to a description of a behaviour pattern. But this pattern has as its aim a way of handling the world and in particular relations with others, mirroring what Mead refers to as 'social intelligence' [6]. Thus roles are created by a society and serve to act as a representation of that society. Consider the words of Epictetus [7]:

Remember that you are an actor in a drama, of such a kind as the author pleases to make it. If short, of a short one; if long, of a long

Figure 10.1 Jean-Paul Sartre. (Reproduced by permission of Mary Evans Picture Library.)

> *one. If it be his pleasure you should act a poor man, a cripple, a governor, or a private person, see that you act it naturally. For this is your business, to act well the character assigned you; to choose it is another's.*

The literary role or fictional character or entity has an added dimension in that there is an underlying literary aim which is paramount. Also, since a character in a novel is made only of the sentences describing him or her [8], there are many linguistic devices which can all add to the formation of the character. Take, for example, the name Murdstone (murder and stony heart) in Dickens or the famous 'ritual gesture of the hands' of Uriah Heep—a name which also provokes disgust [9]. The underlying aims can be quite varied, but all have the common thread of trying to influence people. Whether it be political through a commitment by the author to give a sense of purpose to a work, or just to highlight social problems or even to make us laugh and share other emotions, literature is much more than just communication.

10.3 Scientific entities

At first sight the idea of talking about scientific entities—even superficially theoretical ones—under the same heading as fictional characters seems absurd. The world of science is after all the actual world and not some invented arena. On the other hand, most perceived differences between fictional and scientific entities rely on having clear ideas about truth values in each of these domains, failing which a better understanding of the issues might be gleaned by focusing on the resemblances—science and literature may not be that far apart.

I want to concentrate on theoretical entities although all entities are theory-laden to some degree or another. There are different kinds of entities postulated in science. Some are based on mathematical ideal entities, for example, point masses [10] and are thus not considered real objects in themselves. Others, which usually start life as postulated though unobserved entities, either become more established—as for example electrons—or fall by the wayside—the fate of the aether. When successfully established it is not so much that entities become observed, but that they become enmeshed within the framework of questions and answers which constitute a particular theory. As with Russell, entities are subservient to theories, 'Things are those series of aspects which obey the laws of Physics'[11]. Similarly echoed in Jacob's book *The Logic of Life* [12]:

> *For an object to be accessible to investigation, it is not sufficient just to perceive it. A theory prepared to accommodate it must also exist. In the dialogue between theory and experience, theory always has the first word. It determines the form of the question and thus sets limits to the answer.*

10.4 Fictional and scientific entities: a comparison

In literature the aesthetic function is dominant [13], but it is interesting how some sort of notion of beauty—as applied to theories—in physics has always been in evidence. It goes back to ancient Greek ideas of harmony and natural order. Even in modern times the beauty or simplicity of a theory is considered a commendable feature. Sometimes under the guise of an Ockham's Razor argument there is usually a tacit presupposition that fundamentally nature has to be simple and harmonious. Although, counter to that most scientists would say that faced with ugly facts even the most beautiful theory would go by the wayside. I think that this shows that whereas beauty is not the only consideration in assessing a theory it can certainly be one consideration.

10.5 Beauty

The quest for beauty in theoretical physics, second only to simplicity [14], has always been considered an important factor in the criteria of choice of a theory. It is quite difficult to define precisely what is meant by beauty. Socrates started off by considering that: 'whatever is useful is beautiful' [15], but ended up declaring: 'all that is beautiful is difficult' [16].

In the very first line of his essay on beauty in science, Chandrasekhar took this difficulty as his starting point [17].

Notwithstanding the problems, Chandrasekhar identified two criteria for beauty in science. The first was based on the work of Francis Bacon and invoked the need for an exceptional quality in a theory, beauty being linked to wonder and surprise. The second element formulated by Heisenberg addressed the conformity or self-consistency of a theory to all its parts and as a whole [18].

Chandrasekhar applied his criteria to Einstein's theory of general relativity which scored well on both points. The aesthetic nature of the theory itself or what Jardine refers to as a 'Type 1' aesthetic appraisal is important in the early acceptance of a theory [19]. (For Jardine aesthetic values of the phenomena of a theory and their representations are labelled respectively Types 2 and 3.)

The question is that however one refers to beauty, whether it be utility, self-consistency, wonder, awe, surprise, perfection, symmetry, or a combination of one or more (or all) of these, does it have a place in science? One could answer in the negative, but aesthetic quality seems to be an inescapable part of science. The alternative is to explain what kind of abstraction it is about a scientific theory to which we can apply aesthetic principles and about which we can comment positively or otherwise on the acceptability of a theory.

Beauty does seem to cover, at least in part, the criteria we apply to theoretical constructs. From a structural point of view this fits in very neatly (beautifully?) with the zetetic framework which I have mapped out. One could say that the graph-theoretic shape, which represents the abstract combinatorial form of a theory, is the visual embodiment to which the term 'beauty' applies. We go beyond pure description here. The cognitive layer represented by the graphical form of a theory is a necessary link to the aesthetic form of beauty which many consider important. Without the link it is difficult to explain in general terms precisely what makes theories beautiful.

10.6 Scientific metaphor

To say that science works by the use of metaphors rather than deductions is a truism [20]. But there is much more to the use of metaphor than as a fundamental mode of understanding [21]. Metaphors can be used as weapons in the scientist's arsenal to promote and advance a theory or even as a smokescreen to hide underlying faults which a theory may have. To quote Nietzsche [22]:

Every opinion is also a hiding-place, every word also a mask.

The researcher, as scientific writer, wants to create in the minds of the readership the appropriate reaction. The objective-sounding nature of the style of scientific writing itself is used to accentuate this effect. The mask of detachment between the author and the text is another weapon to be employed. Everything is appropriated to achieve the desired effect: titles, sub-titles, diagrams, headings, to name just a few items from a long list [23].

Even linguistic devices can sometime be important in science. Particularly in names of objects like quarks, gluons and quasars [24]. The choice of names in these cases is quite deliberate and designed to carry a specific message in their use. Even the employment of terms like charm by particle physicists, seemingly arbitrary and playful, conveys an underlying message that the theory itself has this commendable property. Even though the purpose behind all these devices may not be so much aesthetic but evangelical, nevertheless it is often the case that terminology which is displeasing to the ear tends to fall into disuse or is corrupted into an easier-sounding word, for example, quasar instead of quasi-stellar radio source.

It also goes well beyond the written text. Gifford [25] illustrated the 1988 United States presidential campaign between Dukakis and Bush with television pictures of Dukakis in the 'driving seat' of an M-1 tank and Bush waving from a motor launch in a polluted Boston harbour—Dukakis, as governor of Massachusetts, should have got rid of the pollution. Such visual metaphor tyrannically exploited on television is expected in the field of politics; is science so far away from this arena?

I remember once going to a seminar given by a visiting professor whose ideas were generally considered extremely radical and contrary to many in the audience. The dress for these occasions was quite informal, but the speaker wore a dark suit and tie—the only person in the room formally attired. 'How can you not take seriously someone who looks like a bank manager!' was what one of the speaker's followers said to a group of us after the talk. There was laughter at this tongue-in-cheek remark, but the visual rhetorical trick was nevertheless well played.

But in science metaphor is not necessarily a bad thing. As with all tools it can be useful as well as harmful. The main use is to aid understanding by linking ideas and explanations with something more familiar. This is particularly a problem in science where so many theories involve concepts which are far removed from everyday experience. Without some sort of metaphorical framework it would be almost impossible to explain and describe scientific theories— certainly this could not be done by simply quoting equations.

Another benefit from metaphorical networking is to situate ideas within a context. This is both in the sense of bringing ideas and explanations within a family grouping and also to generate new ideas. Photons of light are neither particles nor waves, but it is part of the understanding of the theoretical concept of light that it is in a sense both particle and wave. The fictional idealization or 'as if' world of the unreal is just as important as the world of the so-called real [26].

10.7 Explanation and understanding

According to van Fraassen, a scientific explanation is an answer to a why-question [27]. Salient factors which cause an event thereby explain it. Salmon has argued that [28]:

> ... *not all why-questions are requests for scientific explanations and not all requests for scientific explanation are made by posing why-questions.*

Furthermore, it is not sufficient for an explanation merely to fit events into regular patterns, such empirical-type structures have little explanatory power [29], the lack of causal relationships being the culprit. However, causality is not unproblematic itself, particularly in regard to quantum-mechanical phenomena [30].

One may cast the net wider and include understanding. This goes beyond the causal nexus of events to a unity of 'underlying mechanisms upon which we depend for explanation' [31]. In this way explanations contribute to an understanding which also depends on how the explanations fit in with each other as a whole. At heart there is the problematic experience for us as 'primarily questioning, explanation-seeking, and understanding creatures' [32].

The present zetetic analysis puts questions at the heart of the network rather than events. Answers (and explanations) are thus freed from the confines of being described in purely erotetic terms and together with the questions form the abstracted structure of a theory.

So what we end up having to contend with is the false idea that science is objective, value-free and is able to give the whole picture [33], and the equally false idea that science and literature are fundamentally different pursuits. Behind this is the problematic idea of truth in science and literature. As soon as the idea of an absolute truth is removed from the centre-stage position and replaced by a concept of truth relative to some theory, then literary and scientific truth may converge towards one another [34]. The effect of literature—and at heart it is this effect which is the fundamental measure of literature—is to mould the way we 'perceive and speak of the world' [35]. What literature works on is human perception, through the medium of a fictional world or framework. This fictional world is not just an alternative to the real world, but is created in the mind of the subject through the text. Truth is thus not propositional but conceptual. So the different hierarchical levels of fictional narrative, together with characterizations and presuppositions driven by the conceptual ideas of authors, predispose both a sharing of and a manipulation of concepts in others. The persuasive or rhetorical element therefore is always present to some degree however small. This element is important, though, for it also introduces an interrogative element into literature.

10.8 Interrogative aspects of literature

There are many facets to the questioning or problematic modes in literature. In Roland Barthes' autobiography he reflects [36]:

I do not say: 'I am going to describe myself' but: 'I am writing a text, and I call it R.B.'.

The self has become an enigma, and the role as author differentiated from that of subject. Barthes has in his autobiography reversed the normal role of literary characterization by creating a fictional character out of a real one—himself—instead of trying to create a real character by way of a fiction.

Many texts use the 'absurd' as a means of conveying the problematic nature of the meaning of the texts themselves. Meyer [37] employs an example from Kafka, called 'the test'. In this example an applicant for a servant's position is unable to answer any of the questions at his interview. As he is about to leave, he is told that he has in fact succeeded in passing the test, since those that attempt an answer to any questions immediately fail.

In another example, Sartre considers two lines from a poem by Rimbaud:

Oh seasons! oh castles! What soul is faultless?

The question is seen here as a rhetorical device rather than a genuine question. Even in everyday spoken language rhetorical questions such as: 'Do you think I was born yesterday?' or 'What do you take me for?' have the interrogative form in order to create a special effect. In these cases a reactive tension exists between parties. Patricia Deduck describes Robbe-Grillet's view of the function of art as [38]:

... the creation of an equation balanced by questions which in turn create, then compel, certain answers ad infinitum.

Thus literature, by virtue of its rhetorical aspects, through its modes of writing, basic motivation and direct message, gives expression to problem situations—theatre becomes an 'interrogative act' [39]. In science one judges a theory by the questions it solves, but a text 'by the question it evokes or compels one to ask' [40].

One cannot, of course, sum up literature and science in just one pithy sentence. However, my aim is much more particular and therefore at a very deep and general level certain characteristics of literature and science can be brought out. I am more interested in the grounding of these pursuits and, consistent with the zetetic model, look at the underlying questions posed. The scientific aspect—if I can refer to it in that way—concentrates on the formal aspects of the relations between questions directed at the world of nature. The literary aspect is that of the posing of questions and the creation of worlds of concepts.

Literature is not so much concerned with formal relationships between questions but with provoking, persuading and sharing conceptual ideas which are problematical. At heart, it is this interrogative aspect which links literature and science, and more generally all human intellectual activity.

Notes and references

[1] A version of this section appeared as Sanitt N 1994 The Mary Poppins effect *Phil. Now* No 9 11–2

[2] Sartre J-P 1969 *Kean: Three Plays* (Baltimore, MD: Penguin)

[3] Sartre J-P 1969 *Kean: Three Plays* (Baltimore, MD: Penguin) p 102

[4] *Blood Brothers* a musical and book. Music and lyrics by Willy Russell.

[5] Interview on *Wogan* November 1990 BBC TV

[6] Mead G H 1962 *Mind, Self, and Society* (Chicago, IL: University of Chicago Press) p 141

[7] Epictetus 1966 *Moral Discourses Enchiridion and Fragments: Enchiridion XVII* (London: Everyman) p 260. Note that Epictetus is referring to God when he says 'to choose it is another's'.

[8] Welleck R and Warren A 1949 *Theory of Literature* (Baltimore, MD: Penguin) p 25

[9] Welleck R and Warren A 1949 *Theory of Literature* (Baltimore, MD: Penguin) p 219

[10] Parsons T 1980 *Nonexistent Objects* (New Haven, CT: Yale University Press) p 228

[11] Russell B 1914 *Our Knowledge of the External World* (Chicago: Open Court) p 110

[12] Jacob F 1989 *The Logic of Life* (Baltimore, MD: Penguin) p 15

[13] Welleck R and Warren A 1949 *Theory of Literature* (Baltimore, MD: Penguin)

[14] Heisenberg lists 'simplicity and beauty' as leading to truth, quoted in: Chandrasekhar S 1987 Beauty and the quest for beauty in science *Truth and Beauty* (Chicago, IL: University of Chicago Press) p 65. Ellis lists simplicity and beauty as the first two criteria of choices for theories: Ellis G F R 1993 *Before the Beginning* (London: Boyars/Bowerdean) p 15

[15] Plato *Greater Hippias 295c (The Collected Dialogues)* ed E Hamilton and H Cairns (Princeton, NJ: Princeton University Press) p 1548

[16] Plato *Greater Hippias 304e (The Collected Dialogues)* ed E Hamilton and H Cairns (Princeton, NJ: Princeton University Press) p 1559

[17] Chandrasekhar S 1987 Beauty and the quest for beauty in science *Truth and Beauty* (Chicago, IL: University of Chicago Press) p 59

[18] Chandrasekhar S 1987 Beauty and the quest for beauty in science *Truth and Beauty* (Chicago, IL: University of Chicago Press) p 70

[19] Jardine N 1991 *The Scenes of Inquiry* (Oxford: Clarendon) p 208
[20] Arthur W B 1992 quoted in Mitchell Waldrop M 1992 *Complexity* (New York: Touchstone) p 327
[21] Murdoch I 1992 *Metaphysics as a Guide to Morals* (London: Chatto and Windus) p 305
[22] Nietzsche F 1990 *Beyond Good and Evil* (Baltimore, MD: Penguin) section 289 p 216
[23] Laszlo P 1993 *La Vulgarisation Scientifique; Que Sais-je?* (Paris: Presses Université de France) p 58
[24] Quarks—these are hypothetical building blocks of subatomic matter which have charges in one-third multiples of the electronic charge. The word came from *Finnegan's Wake* by James Joyce.
Gluons—these are particles which 'glue' together quarks to form particles.
Quasars—short for quasi-stellar objects.
Charm—name given to quantum numbers in particle physics.
[25] Gifford D 1991 *The Farther Shore* (New York: Vintage) p 39
[26] Vaihinger H 1968 *The Philosophy of 'As If'* (London: Routledge and Kegan Paul) p xlvii
See also Ogden C K 1932 *Bentham's Theory of Fictions* (London: Kegan Paul, Trench, Trübner) p xxxi
[27] van Fraassen B C 1980 *The Scientific Image* (Oxford: Clarendon) p 134
[28] Salmon W C 1984 *Scientific Explanation and the Causal Structure of the World* (Princeton, NJ: Princeton University Press) p 101
[29] Salmon W C 1984 *Scientific Explanation and the Causal Structure of the World* (Princeton, NJ: Princeton University Press) p 121
[30] Salmon W C 1984 *Scientific Explanation and the Causal Structure of the World* (Princeton, NJ: Princeton University Press) p 254
[31] Salmon W C 1984 *Scientific Explanation and the Causal Structure of the World* (Princeton, NJ: Princeton University Press) p 276
[32] Moravcsik J M 1990 *Thought and Language* (London: Routledge) p 216
[33] Livingston P 1988 *Literary Knowledge* (Ithaca, NY: Cornell University Press) p 107
[34] Rantala V and Wiesenthal L 1989 The worlds of fiction and the worlds of science *Synthese* **78** 53–86
[35] Cebik L B 1984 *Fictional Narrative and Truth* (Lanham, MD: University Press of America) p 215
[36] Barthes R 1988 *Roland Barthes by Roland Barthes* (Engl. Transl. R Howard) (New York: Macmillan) p 56
[37] Meyer M 1986 Problematology and rhetoric *Practical Reasoning in Human Affairs* ed J C Golden and J J Pilotta (Dordrecht: Reidel) p 127
[38] Deduck P A 1982 *Realism, reality and the fictional theory of Alain Robbe-Grillet and Anais Nin* (Lanham, MD: University Press of America) p 48
[39] McGlynn F 1990 Postmodernism and theater *Continental Philosophy III, Postmodernism—Philosophy and The Arts* ed H J Silverman (London: Routledge) p 149

[40] Meyer M 1986 Problematology and rhetoric *Practical Reasoning in Human Affairs* ed J C Golden and J J Pilotta (Dordrecht: Reidel) p 131

Chapter 11

Overview

Any scientific theory must stand or fall on its own merits. However, the consensus view is based on a sceptical rationality by which theories are continually tested against each other and the world. What emerges from this process is not the truth or an absolute objective knowledge of the world, but a changing network of ideas which we call scientific theories. The rationale behind the present work is not only to put forward a model which may be used to represent the whole process of scientific theorizing, but to encourage and advocate a view of science—and theoretical physics in particular—which is somewhat different to the historical norm.

There is an old adage that philosophers who disagree about everything, agree on basics. A corollary to this might be that philosophers who disagree on basics might appear to agree on a number of things. I think that this latter aphorism is quite useful as it refers to a superficial agreement which masks an underlying mismatch of basic views. The superficial agreement is formed merely by a generally accepted view that a particular theory is valid, while there may be different opinions as to how science itself is grounded.

The zetetic model begins from a basic attitude to science which emphasizes the process of science through the activity of its practitioners, and is based on question generation rather than problem solving. This basis, deeply grounded in the ancient ideas of invariance and the questioning approach to the world, reflects the way in which the human mind has developed. Just as there may be basic structures to language which arise out of the evolutionary development of the human brain, so the way that we cope with, live in, and theorize about the world may also rely on similar brain structures. The process of science is not seen as an autonomous object but as a human endeavour of a particular type. I think that in this case it is liberating that theoretical science is not trapped as a rootless esoteric pursuit unconnected with or barely connected to mankind. Instead, it is firmly rooted within the family of intellectual exploration, reflecting a basic instinct to understand nature.

My aim is to view science itself as a scientific problem which has to be resolved, worked upon, and confronted in much the same way that any other scientific topic of research would be handled. Considering that my background is that of research in astrophysics, this attitude is not surprising but, from the theoretical or reflective point of view, there clearly seems to be a paradox. I refer here to the possible generation of an infinite regress. How does one justify a method of approach to a problem, if the problem consists of the justification of the method itself? Needless to say, I diverge from most of my scientific colleagues in that concerns about methodology are furthest from their thoughts when embarking on scientific research. This is not to say that these problems are ignored, but they are somehow implicit in research work and form part of the general background, which comes more into play at the assessment stage, after research is completed and published. My resolution of the apparent paradox is not to claim justification or knowledge of truth but just to set out the arguments, deal with the problems and put forward the theses. In this respect my divergence from my scientific colleagues is clearly illusory. I do not mean to indulge in sophistry, but wish to make the point that science and its philosophy are inextricably linked together.

My starting point is an age-old methodological device in scientific research, namely, finding an entry or handle into a problem. Whenever areas of research prove fruitful it almost always turns out that there were a number of different avenues or handles which might have led to a similar goal. Without the benefit of belated hindsight, however, the fortress which seems to surround a problem can appear impregnable.

However, the problems in research start before even the first volley is loosed upon the castle walls. I refer here to what one can loosely describe as presuppositions or more forcefully as prejudices. Under the guise of objectivity scientists try to eradicate what is essentially the unerasable. I can borrow and transform Descartes' famous phrase: 'I think therefore I am biased'.

Granted that it is impossible to rid our minds of all ideas before we start doing research, how can we proceed? The scientific answer to this question is that we just do, and this is a metaphor for the whole of life not just scientific research. I do not just mean the life of an individual but the whole of life. One cannot remain mute and immobile all the time, even though this would eliminate the problem of presuppositions. Living things by definition cannot be totally inactive.

Thus weighed down by a weight of prejudices how can we ever breach a problem? The first handle that might come to mind is truth, or more precisely, scientific truth. I reject this notion because scientific propositions are fundamentally undefinable. Since whatever one's idea about truth is, it must apply to a statement, it must be unambiguous and it must be absolute.

By unambiguous I mean that a statement must be exclusively true or false, and by absolute I mean that the truth value of a statement can never change. Those statements which satisfy these conditions are termed propositions, and even though some statements may give the appearance of being scientific propositions, they can never satisfy the above criteria. The reason is that not only does science change, theories come and go but even scientific definitions themselves change. 'The Sun will rise tomorrow' may appear to be a scientific proposition, but what do we mean by Sun? If the Sun were to explode into a supernova and engulf the Earth tonight, assuming we could still observe the situation tomorrow, how would we define 'the Sun'? Would we include the engulfed Earth and inner planets? Would we include the supersonic shell of material blown off? Would the Earth's rotation be affected? How would we define 'rise'—or even 'tomorrow'? Clearly, not only is truth not an avenue into the problem, but it cannot even be part of any solution to the problem.

The first part of the handle which I employ is the idea of invariance. It is the perfect handle to any scientific problem because instead of looking at an expansive sea of change in any complex problem, one only has to find that which does not change—one unvarying reference point is all that one needs.

The second part of the handle is to decide exactly what it is that is invariant. The first candidate that I considered—and rejected—was the 'event'. The reason for rejection is similar to the reason why the notion of scientific truth can be rejected. It is clear from the example which I used about the Sun rising that events are not invariant. They are heavily theory-laden and their invariance would imply that scientific propositions were definable and hence that truth was part of science. The second candidate for an invariant quantity was the 'explanation' or 'answer'. This is probably the worst invariant quantity which one could imagine. Rare enough at most times, acceptable answers are always seen as transitory, and at best only a temporary stop-gap between the ghosts of theories past and future.

The third and successful candidate for invariance is the question or problem situation. At first sight this might appear a strange choice because a scientific question is a creation in the mind of a person or group. Granted that it is directed at or about something in the world, then surely this is where we should look for a foundational basis, rather than inside a scientist's head. I am not saying that phenomenology is all that there is to science, but the scientific question is, in my view, the best starting point.

Questions have a number of characteristics which make them ideal as the foundational basis for science, the most important being that questions have no truth value. You cannot say that a question is of itself true or false; questions are not propositions. Even though this is a negative characteristic, it means that we are not in danger of bringing the notion of truth into science, by introducing questions as a fundamental unit.

The invariance of questions refers to the conceptual, problematic situation which a question represents. Theories may come and go, answers to questions may change and theoretical concepts can be redefined, but the questions remain. Once a question is posed and a problematic situation thereby identified, what happens next? You can answer a question by putting it into the context of a network, or be baulked if no theory or answer is available. Alternatively, you can class a question as not being relevant to a particular set of queries under consideration. The one thing that you cannot do though is un-ask a question.

The final part of the 'handle' which I have introduced is the notion of relationship. Since I am placing questions at the heart of scientific theorizing, I need to look at the way questions relate to each other and to the world. We tend to think of going from questions to answers but answers are not stable or invariant quantities. They serve—temporarily—to relate questions to each other, and it is this network of related questions which forms the backbone to what I refer to as zetetic analysis. What we end up with is a view of questions and answers which is a reverse of the usual common-sense picture. Rather than questions leading to answers we have questions leading to questions, and so questions arise out of answers to *preceding* questions. In this scheme it is answers which lead to questions rather than the other way around. The answer to a scientific question is always problematic and must always give rise to another question. There are no bare answers as such, but only answers which lead to other questions. This is fully consistent with the absence of truth or deductive reasoning in the structure of science. This is not to say that deductive or mathematical reasoning is not important in science, only that it is not a visible part of the basic structure.

So what is the structure of science if it is not based on propositional logic? My answer is that it is based on a relational structure between scientific questions, i.e. a combinatorial structure. Such structural schemes have been around since ancient times, but are represented today by what is known as graph theory. In particular, the structure relevant to science is that of an acyclic directed graph. Graph theory provides the mathematical basis to the study of relational structures in general. It is a very rich field of study and for the present purposes provides not only the formal basis for the description of theoretical structures of science, but also introduces concepts which can be usefully applied. The three most important concepts being isomorphism, cycle and hierarchy.

Isomorphism corresponds to the notion of equality of structure. The power of a theory is in the underlying relational structure of its questions, and two structures are the 'same' if they are isomorphic.

The second concept of importance is that of cycle. In any structure of scientific questions the relationship between questions is directional, hence the need for directed graphs. Furthermore, since deductive or propositional structures are excluded, tautologies or cycles are also forbidden. You cannot go round in circles from question to question arriving back at the original question in a loop. We can thus utilize the ready-made concept of acyclicity in digraphs to ensure that tautologies do not appear.

The third concept of hierarchy manifests itself as a rigorously defined, topological characteristic of acyclic digraphs. The questions represented by vertices in a digraph can be numbered and uniquely grouped into levels. The hierarchical nature of scientific theories thus emerges naturally as a mathematical property of the structure.

In addition to providing mathematical concepts which enrich the study of scientific theory structures, graph theory also admits the reverse procedure of mathematically underpinning scientific concepts.

In the present 'reverse' way of looking at scientific questions and answers, it is answers which always lead to questions. The exception to this appears to be questions without antecedent answers, i.e. bare questions or in graph theoretical terms questions represented by vertices with in-degree equal to zero. These questions which I term 'empirical' have antecedent answers—in fact a whole network of questions and answers—which are suppressed. This suppression corresponds to the concept of black-boxing. These are theories or parts of theories which are so well established that they are no longer considered problematical. Once this has occurred then a black-boxed theory becomes equivalent to a deduction and so is rendered invisible on the present scheme. It is always possible though, in the future, for a black-boxed theory to be opened up, if its acceptance is ever questioned; unless and until this occurs, however, such a theory remains closed.

We can thus reiterate the 'cognitive' rules which establish the zetetic framework as follows.

(1) Theory structures are represented by acyclic digraphs of questions (vertices) and answers (arcs).
(2) All free-end vertices have arcs directed away from the free-end.
(3) There has to be a surplus of vertices having in-degree zero over the total of all other vertices.

The third rule ties in three things. Firstly, the difference between empirical and other questions is seen purely in terms of the visibility of the network of questions and answers which give rise to or are antecedent to a question. In this way the distinction between theoretical and empirical questions is intrinsically unimportant; it is more how they are viewed than how they are in themselves. Secondly, what constitutes an acceptable theory is a characteristic of the structure and this structure is built by scientists. Finally, theories are dynamic as well as static; theory change and comparison can be studied in terms of underlying structural comparisons.

Let me say that there is a great deal more to science than theory structures and zetetic structure as described. The rules that lead to the acyclic digraph model are only a means of abstracting a formal structure which is the basis of scientific theories. This formal structure is useful and is amenable to mathematical analysis but it is only a small part of the whole. Having established the framework, the question now is how is zetetic analysis actually applied in practice?

In any area of science, whether it be within an established theory or at the cutting edge of research in new and uncertain domains, the one common characteristic is the lack of clear scientific questions. My point is that when theories do start to emerge the questions or problems and their putative solutions become more well defined. This process of theory building is where zetetic analysis comes into play. There is a translation of a particular problematic area into a network of questions and answers, zetetic analysis thereby provides a cognitive pattern of a theory.

The pattern of a theory or group of theories can also be mapped out by the use of scientometrics. This particular area of bibliometric analysis of science follows the dynamical change and evolution of key areas of scientific research by statistical methods of word-pattern associations in scientific publications. Zetetic analysis provides a theoretical framework which underlies this type of statistical analysis. There is an interplay between the cognitive structures which one would expect and the writings of scientists which are amenable to bibliometric analysis.

There is also an evolutionary angle to science which the present analysis brings out from a different perspective. I do not refer to the somewhat simplistic analogies of 'survival of the fittest' applied to scientific theories, but to a much deeper human—or in a reduced sense biological—trait of pursuing science as part of our nature. In this respect, scientific theorizing is seen as part of a spectrum of cognitive structure, which in one form or another is present throughout the living world. Part of this process is the exploratory behaviour or curiosity which all animals seem to exhibit in various degrees. Such behaviour is in a reduced sense part of a cognitive structure. At the more intellectual end of the spectrum, language itself and other 'structuralist' areas, such as anthropology and the development of intelligence in children, share a common natural genesis.

I have tried to give a number of examples of zetetic analysis applied in practice to various areas of theoretical science. My aim has been more representative than definitive. In early quantum mechanics and relativity we have prime examples of theories which are black-boxed. That is not to say that these theories might not face challenges in the future, but that at present they are considered acceptable. This mature state is easier to analyse because the structure of the theories is so well defined; this has come about by virtue of the theories being refined and redefined in a process which systematically rewrites the past [1]. The present structure of these theories, though intimately based on discoveries in past times, does not maintain the same original chronological sequence.

At the other extreme, I have chosen examples of theories which are problematic and which are more prone to change. (I suspect that by the time I finish writing this work they will have changed quite a lot.) I have also chosen as an example Darwin's theory, partly because its status as a theory has been studied a great deal and also because it is not part of theoretical physics.

The present method of analysis shows quite clearly both the scientific structure of a successful theory and also the problems with the structure of a more problematic theory, such as Dawkins' theory of the extended phenotype. The question is, does zetetic analysis aid understanding by providing explanations in the sort of unified way that is the hallmark of successful science?

Structure of itself does not guarantee explanatory power. 'The train is always late' is an unsatisfactory explanation of why the train is late today [2]. However, by furnishing a network of questions or problems, zetetic analysis highlights the unifying nature of science as well as providing a basis for description. The unification goes well beyond a particular domain of science but encompasses other intellectual domains within nature.

What kinds of criteria are there which determine whether a question or answer is a satisfactory candidate for its role? The answer to this is that there are no objective criteria which one can exhaustively list, much in the same way that criteria for the definition of truth are equally undefinable. What I mean by this is that what constitutes a question—the problematic situation—is defined by the scientist/observer/actor. The answers are also based on experience, experiment, observations and theories but are articulated and defined by the same agents.

Agents may disagree and even within a particular field of science, the structure may look different to different scientists or groups of scientists. Not only is this the case, but any model of science must provide for this to be the case. There are no true answers and even within a particular area, groups may disagree on the relevance or weighting of questions in a network. This is part of the dynamic of science, and the zetetic model provides a basis for the representation and manipulation of this dynamic.

So who decides? The answer here is the same as it has always been: the rational group. This rare species with many pretenders is the social, political and scientific consensus whose interaction both in nature and by nature forms the patterns of scientific thought.

The criss-crossing web of questions, which comprises the constructs of theoretical physics, becomes the object of logical analysis. However, this is not a logic of exclusively true or false propositions, as in mathematical logic, but of combinatorially valid graphs. These simple diagrams are almost to philosophy of science what Feynman diagrams are to particle physics [3]. By constructing the zetetic graph of a theory, one can see the validity of the structure of the theory in terms of whether its graph conforms to the requisite rules. The validation refers, of course, to the structure and not to the details of the theory itself, since the method does not imply a truth value to a theory. In this way, an element of formal mathematics has been introduced into the inductive process which up till now has resisted such attempts. 'Truth' is therefore in the structure not the content. What we have then is a hierarchical scheme for theoretical physics, which adds it to the long list of applications in nature of hierarchical systems [4].

The subjective–objective debate, in terms of the present scenario, becomes a non-issue. Instead, the deductive–inductive aspects of theoretical physics are reflected in the contrast between structure and content: the content of the theories being the questions and answers, and the structure being the network that these form. Similarly, the empirical–theoretical division takes on a different meaning. Empirical questions reflect the acceptance of scientists to a black-boxing of a cluster of questions and answers. The whole question of commensurability of theories and theory change can be viewed much more clearly and neatly in terms of the zetetic scenario and the changing structures which are accommodated within the model.

Finally, theory comparison becomes less of an art and more of a science, as theory structures can be compared in a much more analytic way within the model. It is surprising that up till now the only attempt to quantify comparisons between theories has been by Peebles and Silk [5]. Their approach was ostensibly that of game theory, where the odds of a number of observable phenomena are calculated against each of a number of rival cosmological theories to see which comes out on top. Not all phenomena considered have equal weight, and Peebles and Silk also ascribe individual probabilities to the success or otherwise of a theory to explain a given observation. In this way each theory (in their 1990 paper they consider five rival theories) is considered in turn as a black-box which operates on the observations, generating probability and hence success. Their tabular presentation does not bring out the cross relations between observational evidence, as each observation is treated independently. However, such an approach as theirs is very much in the same spirit as the present model. In order, I think, to make philosophy of science more palatable to astrophysicists, Peebles and

Silk refer to their readers as 'punters' and put forward their aim, which is to 'enrich, enlighten and even amuse those of our colleagues who are trying to assess the merits of the rival cosmogonies'. I am certainly in favour of such an approach if it encourages scientists to think about the philosophical aspects of their work. Underlying Peebles' and Silk's light-hearted preliminaries however, is a serious attempt to analyse a difficult and confused situation.

By translating a scientific theory into a series of questions and fitting these, with their answers, into a hierarchical framework, the logical structure of the theory unfolds. It is this structure of human manufacture which regenerates further structures in the hierarchy, to form a never-ending chain of reasoning—which is the process of science.

Notes and references

[1] Baudrillard J 1994 *The Illusion of the End* (Cambridge: Polity)

[2] See the entry 'covering law model' in Flew A 1979 *A Dictionary of Philosophy* (London: Pan Books) p 79

[3] Named after R P Feynman who in the late 1940s introduced a pictorial representation for particle processes in quantum electrodynamics. See, for example, J R Aitchison and A J G Hey 1982 *Gauge Theories in Particle Physics* (Bristol: Hilger)

[4] Allen T F H and Starr T B 1982 *Hierarchy* (Chicago, IL: University of Chicago Press)

[5] Peebles P J E and Silk J 1988 *Nature* **335** 601–6; 1990 *Nature* **346** 233–9

Index